A CATHOLIC ENGINEER *Searches* *for* GOD

A Story of Empowerment

JACK CARLOW

Jack Carlow Publisher
P.O. box 555
Hingham, Ma 02043

www.jackcarlow.com

ISBN 13: 978-0-9964586-0-3

Published in the United States by Jack Carlow Publisher.

To those who question...

Contents

Introduction **13**

Part I: Gods, Humans, & Knowledge **15**

Chapter 1: The Evolution of God **16**

Chapter 2: A Very Brief History of Abrahamic Religions **19**
Prehistory (Before Ancient Egypt) 19
Ancient Egypt—3300 BC to 500 BC 20
Ancient Greece—500 BC to 146 BC 22
The Roman Empire—146 BC to 476 AD 23
The Dark Ages—476 AD to 1400 AD 25
The Renaissance/Age of Exploration/Scientific Revolution/ Industrial Revolution—1400 AD Onward 27

Chapter 3: The Egypt of Moses **31**
Daily Life in Ancient Egypt 31
Medicine 36
Religion 37
Akhenaten 40

Chapter 4: Moses and the Bible Writers **42**
Some Things They Observed 42
Some Things That Moses and the Bible Writers Believed or Didn't Understand 43
The Bible 45

Chapter 5: Visible Ignorance vs. Invisible Perfection **50**
Gods That "Talked" 50
The Male God 51
Two-Thousand-Year-Old Knowledge 51

Chapter 6: Gods We Don't Need Anymore **53**

Part II: Proteins & Protons **59**

Chapter 7: Proteins **60**
Life 60
How Did Life Begin on Earth? 65
Evolution 69
Why Can't We Observe Evolution Happening? 72
Is Evolution a Fact or Just a Theory? 74
What Does it All Mean? 75

Coming Soon 75
Life in the Universe 76
Chapter 8: Protons 79
The Big Bang 79
What Caused the Big Bang? 81
The Spread and Acceptance of Knowledge 81

Part III: Time & Distance 85
Chapter 9: Our Perception of the World 86
Our Shrinking World 86
How Big Is a Million? 88
How About a Billion? 89
A Trillion? 89
Chapter 10: Our Neighborhood in the Universe 91
The Solar System 91
Our Star: The Sun 92
Our Planet: Earth 92
Chapter 11: Big and Beyond Big 94
Big: Our Galaxy, the Milky Way 94
Beyond Big: Our Universe 95
How Many Stars Are There? 96
Traveling 96
Intelligent Life 98
Other Universes: The Multiverse 99
What Does This All Mean? 99

Part IV: Our Current God/Religion 101
Chapter 12: The "Ancient Egyptian" Catholic Church 102
Chapter 13: The "Holy" Roman Catholic Church 105
Pagans 105
Heretics 106
Witches 109
Jews 109
The Crusades 110
Religious Wars 113
Native Peoples 114
"Holy"? 116
Chapter 14: Some Criminal Popes (& Infallibility) 117

Chapter 15: The Institute for the Works of Religion 127
Chapter 16: Child Sexual Abuse Cases 130
Chapter 17: The Church's Treatment of Women 132
Chapter 18: Crimes of Other Religions 134
Chapter 19: The Church's Suppression of Knowledge 136

Part V: Do We Need a God? 147
Chapter 20: What If People Didn't Believe in God/Religion? 148
Would Good People Turn Bad? 148
The Religion-Government Complex.................................. 150
Imagine No Religion.. 151

Part VI: Believing.. 159
Chapter 21: Why Do We Believe? 160
Programming.. 160
Tradition... 162
Hope... 163
Inertia 164
Fear 165
Comfort.. 166
Mental Illness... 166
Not Using the Brain/Lack of Courage/Irresponsibility/
Lack of Self-Belief/Ostrichitis 167
Conformity.. 167
Questioning... 168
Could "Belief" in Something That Doesn't Exist Be
Harmful? 169

Part VII: Proof 173
Chapter 22: Talking With God 174
Chapter 23: Santa... 177
Chapter 24: God's Attributes and Inadequacies 179
God's Attributes 179
God's Insecurity 180
Miracles and Prayers.. 180
Chapter 25: Made-Up Stuff.. 184
Virgin Birth 184
Was Jesus God?.. 187

Papal Infallibility 187
Limbo 188
Sanctity of Life 189
Chapter 26: Moses and the Commandments 190
The Bible—Old Testament 191
Illness, Disease, and Death In Moses's Time 192
If You Were Moses's God 194
Moses's God 194
Leprosy & Leviticus 195
The Truth about Moses's God 196
Chapter 27: Birth, Marriage, Death, & Intuition 198
Chapter 28: If 202
Chapter 29: The Greatest Ponzi Scheme of All Time 204
Ponzi Schemes 204
Some Religious Scams 208
Other Foolishness 211
How to Judge a Preacher 212
Jesus and the Vatican 212
Summing Up 213
Chapter 30: Atheists and Agnostics 220
Chapter 31: You're Not Alone 221
Chapter 32: "Love Thy Neighbor" 226
Chapter 33: Good & Evil/Heaven & Hell/Eternity 228
Good & Evil 228
Heaven & Hell 229
Eternity 230
Chapter 34: Purpose 232
The Bad News 232
The Good News 232
Chapter 35: "Deus de Nonsensus" 234
Epilogue 242
Notes 243
Bibliography 261
Acknowledgements 270
About the Author 271

A CATHOLIC ENGINEER *Searches for* GOD

Introduction

Do you believe in God? Most people do, but in the not-too-distant future, possibly you, maybe your kids, and probably your grandkids, won't—nonbelief is rising.

Like most of you, I was raised with religion and God. I was baptized Catholic, confirmed, and graduated from parochial high school.

My Mom and Dad were devout, raised us in the Catholic faith, and expected us to go to church, be respectful, and care for others. (In those days, Moms were the backbone of the church.)

I attended Mass and Sunday school every week (along with Confession on Saturdays), received plenary indulgences for attending First Friday Masses, went to many Christmas Eve midnight Masses (followed by wonderful buffets at friends' houses), and didn't eat meat on Fridays. (What I wouldn't give now for my Mom's baked macaroni, fish cakes, and fish chowder.)

In her last years, my Mom spent much time in a nursing home due to dementia and Parkinson's and at the end felt that her faith and her God had abandoned her—very sad. (On the other hand, my brother tells me that some churches in Florida are quite wealthy, fueled by the elderly eager to donate their money as the end draws near and Heaven hopefully awaits.)

As I grew older, questions and doubts emerged. I didn't go to Mass and ate meat on Fridays. I let things stew in my mind, processing subconsciously, altering ideas, opinions, and beliefs with little active thought. Gradually, I began my investigation of God and religion.

As children, we're taught to believe the teacher, politician, police officer, coach, priest, and Church. As we mature, we begin to examine and question—it's part of growing up, part of human nature. We question, we analyze, we investigate, we read, we discuss. To not do so would be negligent to ourselves and society. How else could we come to a reckoning about God and religion?

So if you've ever wondered, if you've ever been curious about God and religion, join me on this search. Let's see what we discover.

No matter the outcome, we'll be better off for the search, since it will let us confirm or deny the indoctrination we received as children; we'll be empowered.

Part I
Gods, Humans, & Knowledge

Chapter 1

The Evolution of God

Gods wreaked havoc, unleashed chaos and mayhem, and caused the mysterious, unexplainable, and the miraculous long ago. Events, actions, matters, calamities, and things that were beyond human understanding were believed to be done by gods. Throughout history, hopeful and somewhat obedient humans worshipped thousands of different gods and goddesses and formed multitudes of different religions that organized and controlled the believers. These gods eventually evolved into today's one all-powerful God for most believers, and the people of the world now practice the following religions.[1]

Religion	Percentage of the World Population Practicing This Religion
Christianity	32%
Islam	20%
Judaism	Less than 1%
Buddhism	6%
Confucianism/Taoism	6%
Hinduism	13%
Other Religions	6%
Nonreligious	16%

(Buddhism, Confucianism, and Taoism are usually described as "ways of living" as opposed to god-based religions.)

Some Key Events in the Evolution of God

Date	Key Event
100,000 BC	Evidence of ceremonial burials
10,000 BC	Gobekli Tepe religious site in Turkey
3600 BC	Religious buildings in Malta
3200 BC	Newgrange religious/burial site in Ireland
2600 BC	Stonehenge religious/burial site in England
2560 BC	Great pyramid of Egypt—Pharaoh Khufu's burial tomb where his spirit "could ascend to the heavens"
2000 BC–1850 BC	Abraham—patriarch of Judaism, Christianity, and Islam
1500 BC	Hinduism founded
1393 BC–1273 BC	Moses
1344 BC	Pharaoh Akhenaten in Egypt does away with multiple gods and establishes one god—Aten
1300 BC	Judaism founded
500 BC	Confucianism founded
450 BC	Buddhism founded
4 BC–33 AD	Jesus
70 AD	Christianity sect of Judaism founded
380 AD	Christianity becomes the official religion of the Roman Empire
476 AD	Decline of the Roman Empire—beginning of the Dark Ages
622 AD	Islam founded
1095 AD–1291 AD	The Crusades
1299 AD–1870 AD	The Inquisition
1455 AD	Guttenberg makes the first printed Bible
1492 AD	Columbus opens new continents for Christianity

Date	Key Event
1534 AD	Henry VIII breaks from the Roman Catholic Church and establishes the Church of England
1788 AD–1791 AD	The US Constitution and the Bill of Rights grant the freedom of religion to the people
1789 AD–1799 AD	The French Revolution demotes the political power of the Church

Why haven't new religions developed and replaced these older religions? New sects of Christianity, like Mormonism and Christian Science, have developed, but for the most part, tradition, government endorsement of these older religions, and the rise of knowledge have limited the opportunity for major new religions to catch on.

Which of these religions is the "one true religion"? Could all of them be wrong?

Let's take a closer look at the three Abrahamic religions (Judaism, Christianity, and Islam), which are practiced by more than half the world and by 75 percent of all people who practice a god-based religion—with particular emphasis on Christianity.

Chapter 2

A Very Brief History of Abrahamic Religions

Prehistory (Before Ancient Egypt)

Modern humans arose in Africa two hundred thousand years ago and began spreading around the world fifty thousand years later. They used fire for cooking and warmth, made tools, developed language, created art, and practiced religious rituals in burying their dead.[2]

Agriculture developed around 10,000 BC, allowing nomadic hunter-gatherers to settle down in one place, begin farming, and develop large civilized societies. Gradually, civilizations arose in Mesopotamia, Egypt, India, and China around 3500 BC. People lived in villages, cities, and concentrated regions in relative peace and stability; formed central governments; developed economies; and created complex language, writing, and culture. People could now specialize in farming, craft making, trading, soldiering, and so forth. Religions developed and made gods of the sun, moon, earth, sky, and seas accompanied by magical beliefs and superstitions.[2]

Ancient Egypt—3300 BC to 500 BC

Egyptian religious beliefs evolved from prehistoric beliefs and had thousands of gods and goddesses. Angry gods caused natural disasters and sickness; helpful gods cured people, made the Nile flood at the right time of the year to enrich the soil, and made day follow night. (Several gods could have responsibility of the same thing, like the sun.)

A Few Egyptian Gods and Goddesses[3,4,5]

Amen	king of the gods, god of the wind
Ammut	crocodile goddess, devourer of the dead
Anubis	jackal god, god of the dead
Anuket	goddess of the Nile River
Apophis	god of snakes, god of chaos
Aten	the sun god
Babi	god of baboons
Bast	cat goddess, protector of pregnant women and children
Bes	dwarf god, protector of newborn babies
Geb	god of the earth
Hapi	god of the Nile
Hathor	cow goddess, goddess of love
Heget	goddess of frogs, goddess of childbirth
Henet	pelican goddess, protector against snakes
Horus	falcon god, god of war, god of the sky, son of Osiris and Isis
Isis	goddess of magic and healing, wife of Osiris, mother of Horus
Khnum	the ram god, god of the flood
Khonsu	god of the moon, god of healing, son of Amen and Mut
Ma'at	goddess of justice, truth, and order

Menhit	lion goddess, goddess of war
Meretseger	cobra goddess, goddess of the valley of kings
Min	god of sexual prowess
Mut	creator goddess
Neith	goddess of war
Nekhbet	vulture goddess, protector of the pharaoh
Nephthys	hawk goddess, protector of the home
Nut	goddess of the sky, mother of the sun, moon, and the stars
Osiris	god of the underworld and afterlife, husband of Isis, father of Horus
Ptah	god of craftsmen
Ra	the sun god
Sekhmet	lion goddess, goddess of destruction and war
Serqet	scorpion goddess; goddess of magic
Seshat	goddess of writing, mathematics, architecture
Set	god of evil
Sobek	crocodile god
Taweret	hippopotamus goddess, goddess of fertility
Tefnut	cat goddess, goddess of the rain
Thoth	ibis god, god of the moon, wisdom, writing
Wadjet	snake goddess, protector of the pharaoh

The pharaohs were thought to be divine—the representative of the gods—and were the political and spiritual leaders. The priests were the elite of society, and the people appealed to the gods through the priests with offerings, rituals, and prayer. Morality was based on the belief that each person should live an honorable life and not interfere in the lives of others.[6] Religion helped maintain order and peace, but it controlled medicine and science, which were primitive compared to today.

At the time, the land of today's Israel lay between two superpowers: Egypt and Mesopotamia (Syria and Iraq), and the people in this entire area believed in many gods. Abraham (around 2000 BC) and Moses (1393 BC–1273 BC) lived in Egypt, observing and absorbing the culture, knowledge, and traditions. Judaism, Christianity, and Islam all trace their roots to Abraham. Judaism began forming around 2000 BC when God revealed himself to Abraham, and it became more organized seven hundred years later, when God appeared to Moses and gave him the commandments. Christianity began around 70 AD, and Islam, descended from Abraham's son Ishmael, began in 622 AD. All three religions believe in one god (although Christianity believes in the Trinity—three Gods in one) and are, in a sense, cousins.

Ancient Greece—500 BC to 146 BC

Many Greek religious practices can be traced to Egypt, including some of the same gods, like Isis, but the Greek gods were a little more humanlike and less mystical and mysterious.

Since pharaohs ruled Egypt and Greece was democratic, the Greeks were more open to new ideas, discussion, and research. While they had many gods, reason and logic slowly influenced medicine and science.[7] Contrary to Egyptian beliefs, Greeks determined that angry gods did not cause illness and plagues and that the earth was round, not flat.

A Few Greek Gods and Goddesses[8]

Aphrodite	goddess of passion, love, beauty
Apollo	god of the sun
Ariadne	goddess of the moon
Artemis	goddess of nature
Astarte	goddess of fertility

Athena	goddess of war
Circe	goddess of death
Demeter	goddess of the harvest
Dionysus	god of wine, ecstasy, and theater
Eros	god of love, passion, and romance
Gaia	goddess of the earth
Hera	goddess of marriage
Hymen	god of marriage
Muses	goddesses of inspiration
Nemesis	goddess of revenge
Pan	god of nature, laughter, and music
Poseidon	god of the sea
Zeus	the main god, the god of thunder and rain

The Roman Empire—146 BC to 476 AD

When Rome conquered Greece in 146 BC, it took much of the Greek religion and merged it with its own; the Romans also had hundreds of gods.

A Few Roman Gods and Goddesses[9]

Apollo	god of poetry, music
Aurora	goddess of the dawn
Bacchus	god of wine, pleasure
Bubona	goddess of cattle
Ceres	goddess of the harvest
Cupid	god of love
Diana	goddess of the moon, nature
Fama	goddess of fame
Flora	goddess of flowers
Fortuna	goddess of fortune
Hercules	god of strength

Janus	god of beginnings and endings
Juno	queen of the gods
Jupiter	king of the gods, the storm, the air and the sky
Justitia	goddess of justice
Libertas	goddess of freedom
Luna	goddess of the moon
Mars	god of war
Mercury	messenger of the gods
Minerva	goddess of wisdom
Morta	goddess of death
Neptune	god of the sea
Pax	goddess of peace
Saturn	god of the harvest
Securitas	god of security
Sol	the sun god
Tempestas	goddess of storms
Trivia	goddess of magic
Venus	goddess of love
Veritas	goddess of virtue
Victoria	goddess of victory
Voluptas	goddess of pleasure
Vulcan	god of fire

The Romans emphasized public health more than the Greeks or Egyptians did. They were more interested in preventing illness than in developing cures (maybe because they didn't really have any "cures"). They built aqueducts to bring water to cities, and public baths were everywhere and used by the rich and poor alike. They used the seven rivers to flush out the sewers.[10]

Around 70 AD, Christianity split off from Judaism and became its own religion, adopting many features of Egyptian, Greek, and Roman religions (and later, Germanic and Celtic paganism). Then, after a few hundred years of persecution, Emperor Constantine (perhaps being a convert to Christianity or seeing the spread of Christianity and thinking it could help organize and control the people) made Christianity the official Roman religion in 380 AD, abolishing all the other gods. This gave Christianity the foothold it needed to spread throughout the Roman Empire, which spanned from England to Mesopotamia.

Eventually, the costs of maintaining huge armies in remote lands and the growing government bureaucracy and taxes weakened the Roman Empire, making it more susceptible to barbarian invasions. The gradual decline of the Roman Empire over a few centuries coincided with the spread of Christianity.[11]

The Dark Ages—476 AD to 1400 AD

The fall of the Roman Empire began a period known as the Dark Ages that lasted for a thousand years. Much knowledge was lost when the barbarians destroyed libraries, and the development of new knowledge, science, and medicine was stalled. In northern Europe, people still worshipped many gods.

A Few Northern European Gods and Goddesses[12]

Arianrhod	goddess of reincarnation
Brigid	goddess of fertility and poetry
Cernunnos	the horned god
Cerridwen	goddess of the moon and harvest
Freya	goddess of the moon and marriages
Morgan	goddess of water, healing, magic
Odin	god of war, magic, wisdom
Sunna	the sun goddess
Thor	god of the sky and thunder

After Rome fell, independent small states and countries eventually emerged, all associated with the Church, which became a religious and political power. It controlled the development of new knowledge and ideas. If the Church didn't like a person's idea or theory, the person could be deemed a heretic and punished. It was an authoritarian society, much like we see in the Middle East today.

From 1095 to 1291, the Church promoted the Crusades, which were wars against the Muslims in the Holy Lands. The Crusades increased the power and prestige of the Popes but caused the deaths of several million men, women, and children.

In medicine, prayers to the saints and God replaced the spells and prayers from Egyptian, Greek, and Roman religions. Western knowledge was concentrated in monasteries until the development of universities in Italy, France, England, and Germany beginning around 1100 AD. (The Islamic golden age ended in 1258 AD as a result of the Mongol destruction of its libraries and universities.).

In addition to the Crusades, there were frequent wars and much killing. Then, between 1347 AD and 1350 AD, the Black Death (plague) struck, and one-third of the population of Europe died. From Asia through Europe, over a hundred million people succumbed. Religious fervor blamed the plague on strangers, foreigners, and Jews, and thousands of them were murdered.[13]

At the time, medical knowledge had stagnated. The medical school in Paris told the king of France that the heavens had caused the plague.[14] Barbers were in charge of surgery, and the diagnosis of illness was sometimes based on astrology. Bloodletting was a popular treatment as disease was thought to be caused by an excess of blood in the body. A serious disease meant cutting a vein, resulting in a large quantity of blood being lost; a milder condition used leeches.[15]

The Renaissance/Age of Exploration/Scientific Revolution/ Industrial Revolution—1400 AD Onward

Gradually, knowledge and ideas spread, and the authoritarian control of science and knowledge by the Church weakened. Religion and superstition were replaced by reason and knowledge.

The development of the printing press in 1439 allowed a wider spread of information and the faster creation of new ideas and theories.

The great explorers spread European power and Christianity around the globe. Wealth from newly conquered territories flowed to the kings and queens of Europe.

Democracy arose in America and France. Technological progress accelerated. Freedom brought an individual's right to property, the rule of law, and free market economics. Art and science blossomed.

The spread of increasing freedom and weakening of Church control resulted in a knowledge boom that continues to this day. New discoveries and developments accelerate in free societies and knowledge obsoletes gods.

So now we have a list of inventions, scientific discoveries, and new knowledge replacing the list of gods. (Notice the lack of discoveries during the Dark Ages, 476 AD–1400 AD. More on this in chapter 19, "The Church's Suppression of Knowledge." The horse collar, gunpowder, and the mariner's compass were invented in China during this time.)

A Few Knowledge Milestones through the Ages[16, 17, 18, 19]

Date	Development/Discovery/Invention
1.8 million years ago	Fire and cooking
400,000 BC	Spears
60,000 BC	Bow and arrow

Date	**Development/Discovery/Invention**
16,000 BC	Pottery
3500 BC	The wheel
3200 BC	Sailing
3000 BC	Papyrus/Bronze
515 BC	Crane
500 BC	Cast iron
420 BC	Catapult
280 BC	Water wheel
200 BC	Paper
100 BC	Arch bridge and dam
450 AD	Horse collar
560 AD	Building dome
800 AD	Gunpowder
1119 AD	Mariner's compass
1286 AD	Eyeglasses
1439 AD	Printing press
1543 AD	Copernicus: sun at the center of the solar system
1600 AD	Earth's magnetic field
1608 AD	Telescope/microscope
1615 AD	Galileo: Earth revolves around the sun
1665 AD	The cell
1666 AD	Calculus
1675 AD	Microorganisms
1687 AD	Newton's laws of motion/gravity
1700 AD	Piano
1709 AD	Thermometer
1712 AD	Steam engine
1774 AD	Oxygen
1789 AD	Modern chemistry
1798 AD	Vaccination

Date	Development/Discovery/Invention
1803 AD	Atomic theory
1804 AD	Steam locomotive
1818 AD	Blood transfusion
1831 AD	Electromagnetism
1838 AD	Morse code
1842 AD	Anesthesia
1859 AD	Darwin: evolution
1861 AD	Germ theory
1876 AD	Telephone
1878 AD	Cathode ray tube
1879 AD	Light bulb
1882 AD	Electric power grid
1890 AD	Antitoxins
1895 AD	X-rays
1896 AD	Radioactivity
1900 AD	Quantum theory
1903 AD	Airplane
1905 AD	Einstein: special relativity
1908 AD	Automobiles
1915 AD	Einstein: general relativity
1923 AD	Television
1924 AD	Milky Way galaxy
1927 AD	The Big Bang
1928 AD	Penicillin
1937 AD	Jet engine
1938 AD	Nuclear fission
1945 AD	Atomic bomb
1947 AD	Transistor
1951 AD	Nuclear power
1952 AD	Cloning
1953 AD	DNA

Date	**Development/Discovery/Invention**
1954 AD	Kidney transplant
1957 AD	Sputnik
1958 AD	Integrated circuit
1960 AD	Laser
1967 AD	Heart transplant
1969 AD	First man on the moon
1971 AD	MRI
1971 AD	CAT scan
1974 AD	Personal computer
1990 AD	World Wide Web
1995 AD	First planet of a distant star
1998 AD	Stem cells
2001 AD	Human genome
2012 AD	Higgs boson: "god" particle

Chapter 3

The Egypt of Moses

Daily Life in Ancient Egypt

Egypt was the most advanced civilization of its time but, compared to today, knowledge was very limited. There were no books in public circulation, and what little knowledge the average person possessed had been passed down from ancestors, parents, and priests by word of mouth.

What was daily life like? What did people believe? How did the limited knowledge of the time, influence Abraham, Moses and the other Bible writers?

The Population

The population was three million to five million people spread over twenty thousand villages and towns. A large city, like Alexandria, had a population of three hundred thousand people. To us, the people as a group would appear to be youthful; that's because life expectancy at birth was only twenty to thirty years. The death rate of newborns was about one in three, and half of the children did not make it to age five. Women died frequently in childbirth, and funerals were common occurrences.[20, 21, 22]

Famines occurred every few years. Malnutrition caused scurvy, anemia, and other illnesses. Waterborne diseases, tuberculosis, and other infectious illnesses were widespread, and periodically, plagues broke out.[20, 21, 22]

Population growth was kept under control by infant and childhood mortality, death of women during childbirth, prolonged nursing of infants, and women not remarrying after divorce or death of their husbands.[20,21, 22]

Some Key Population Statistics[20, 21, 22]

Infant Mortality, First Year	30%
Child Mortality, Ages 1–5	20%
Life Expectancy at Birth	20–30 years
Average Age of Adult Deaths	30–40 years
Annual Population Growth	Negligible
Illiteracy Rate	Greater than 90%

The Economy

Agriculture was the major portion of Egypt's economy. They grew grains, vegetables, and fruits; raised cattle, goats, pigs, and fowl; caught fish in the Nile and Mediterranean; and grew flax (linen) and cotton for clothing.

Taxes were paid to the state in the form of labor and grain (which could be easily stored); there was no money.[20] Armies, state projects like temple and pyramid building, and the pharaoh's needs and lifestyle were supported by taxes. (Heavy taxes were considered to be one of the contributing factors in Egypt's eventual decline.)[20]

Occupations

Over 90 percent of the people worked the land. And while a few possessed their own land, most worked land that was owned by the state or the temples. Farm workers never traveled far, so their knowledge of the world outside of their towns or villages was very limited.

In addition to farmers, there were bakers, brewers, carpenters, weavers, stonemasons, jewelers, leather workers, potters, shopkeepers, miners, engineers, boatmen, soldiers, scribes, and priests. And there was slavery, which was commonplace in the ancient world.

Young boys learned a craft or trade from their fathers; girls learned from their mothers. For the few who could afford it, boys were sent to school around age seven to learn religion, reading, writing, and arithmetic but not many went.[20] Although they had a writing system (hieroglyphics and a form of cursive writing called hieratic) and papyrus (as of 3000 BC), less than 10 percent of the people were literate.[21,22]

The upper class and administrators were schooled. Administrators usually inherited their jobs from their fathers, making it difficult for the poor to change social status. With no upward mobility, a wide wealth gap existed between the upper class and the average worker.[21, 22]

But social unrest was rare; most were satisfied with their lot in life or were afraid to complain.[20]

Energy

The sun supplied the energy to grow crops, dry mud into bricks for building houses and buildings, and provided a livable temperature. Fire was used for cooking, metalworking, glassmaking, and pottery, and they probably used charcoal or dried vegetation for fuel. There was no coal, oil, natural gas, or electricity. Sesame seed oil and castor oil were burned to provide lighting.

Transportation

Walking was the way to get from here to there. Men carried goods on yokes on their shoulders; women carried goods on their heads or hips. Donkeys were used to carry goods—not for riding—and loads were also dragged on wooden sleds. The wealthy and military used horses for chariots, but they were not used in the economy as they were expensive to maintain. The privileged were carted about on wooden chairs atop donkeys or carried by men.

Deep rivers and canals were crossed by ferry—there were no bridges. Crossing rivers on foot risked contact with crocodiles and hippos. Barges and sailboats moved goods along the Nile.

Building Technology

The Great Pyramid, built in 2560 BC, is the only one of the seven wonders of the ancient world to survive to modern times. Built of two million blocks of limestone (each block weighed 2.5 tons), it took twenty-five thousand workers over twenty years to build. The blocks were quarried several miles away, floated on barges, and dragged on sleds and rollers to the construction site.[21]

The pharaoh, being the absolute ruler, was able to organize and compel the people to build the pyramids as well as temples, obelisks, and tombs. The pyramid engineers used simple devices like the lever, roller, and inclined plane, and while these projects were massive in size, they were not technologically complicated. They lacked pulleys and hoists or cranes and relied on human muscle power to pull the blocks of limestone up the inclines. (Many historians feel that the Egypt's construction capabilities have been vastly overrated.)[23]

Farming

The Nile Valley was so fertile and manpower so abundant that agricultural techniques remained unchanged for thousands of years. While cattle helped plow fields, the plows were fastened inefficiently by ropes to the cattle's horns; harnesses were poorly designed. Other farming tools were primitive—mostly made of wood and stone.[24]

Metal Working

They used copper, gold, iron, bronze, and tin, and were skilled artisans and jewelry makers.

Science

Religion ruled science, limiting their ability to understand and explain natural events. They believed that the earth was a flat-shaped disc floating on water with Heaven above and the underworld below.[25,26] They thought that the sun and stars revolved around the earth and that the earth was at the center of the universe.

Hygiene

The floors of their homes were dirt. There was no indoor plumbing, and water was used sparingly due to the amount of work involved raising it from public wells or the river and transporting it to their homes. They used washbasins, and since they often walked barefoot, some had wooden or clay foot basins in their homes. They bathed in the river or canals. They shaved their heads and bodies to eliminate lice, and men and women used eye makeup, perhaps to help keep dirt and sand out of their eyes.[20] They disposed of waste in the fields or canals; there were no sewerage systems.

Food

The workers were paid by their employers with beer, bread, oil, and vegetables. The standard wage for a low-paid worker was ten loaves of bread and two jugs of beer each day. (Higher-paid officials had an accounting method to "save" or credit their wages for later use or barter.)[27]

Staple foods were beer, bread, onions, vegetables, and dried fish. Meat was not eaten often as it was expensive. Beer (less alcoholic than modern beer) was safer than water drawn from the polluted river or canals. Milk was rare; there was no pasteurization or refrigeration—in the heat, it spoiled in hours. Many temples had vineyards that made wine for rituals and use of the priests, and the people used wine on festive occasions.

Medicine

The priests were also the doctors and believed that gods, demons, and evil spirits caused illness and disease and, therefore, the cures were prayer, spells, and rituals. Patients would wear magical charms and amulets to drive out evil spirits.

Primitive "medicines," such as herbs, were expected to ease a person's pain, and magic caused the cure. (Opium was used for severe pain.) Some "medicines" contained animal dung, which could have caused serious infection, leading to death. Other harmful ingredients included flyspecks, lizard blood, and swine teeth.[28, 29, 30, 31]

Smallpox, diarrhea, dysentery, typhoid fever, and jaundice caused many deaths, as did malaria, pneumonia, and tuberculosis. Minor infections could result in death. Intestinal pains and headaches probably were untreated. The remedy for a head cold was to drink the milk of a mother who had recently given birth to a boy.[31]

Their restricted diets led to widespread malnutrition. Men possibly reached a height of five and a half feet; women less. Vitamin deficiencies led to poor oral hygiene, and cavities and abscesses were common. People died from infected teeth that weren't extracted.[28, 29]

They could set broken bones and somewhat treat burns and could stitch wounds but surgery inside the body was impossible. Bandages were made of willow leaves. While they understood that the pulse and heart were related, they had no understanding of the circulation of blood. And they believed that thinking occurred in the heart.[29, 30, 31]

A test for fertility was to tap a woman with the tip of a finger on her shoulder—if she twitched, she would bear a child. They believed that a fetus's white parts were formed by semen and colored parts from menstrual blood.[29, 30, 31] They knew how to mummify the dead. Cleanliness was promoted for religious reasons, not for health. Most people were dead by age forty.[28]

Religion

The belief in gods and the afterlife began thousands of years before Egypt.[32] The sun has been the object of worship of many cultures throughout history. It creates day and upon leaving, causes night. Its warmth and light makes all life possible. If you lived then, it would have seemed very logical to believe in Ra, the sun god. Solar eclipses were probably magical and terrifying, and comets were thought to be evil omens.

Since the principal gods were heavenly bodies (the sun, the moon, the stars), the priests paid great attention to calculating when these "gods" (heavenly bodies) would appear, leading to the development of a calendar that allowed the priests to "predict" the best times to plant the fields and when the Nile would flood. This priestly "magical" ability probably enhanced their power and influence over the people.[32]

Thousands of local gods protected the villages against evil spirits and demons who brought illness, disease, infertility, and natural disasters. Some local gods, like Isis, achieved national followings and even became important in other empires, such as Greece and Rome. Gods could merge with other gods, as Ra (the sun god) did with Amen (god of the wind), becoming Amen-Ra, while still maintaining their main duties (the wind and the sun).[32]

A god's statue was paraded through the streets with joy, dancing, singing, and wine during holiday festivals[20] (as happens today in certain communities). One festival honored Osiris, who had died and was resurrected, instilling in the people hope for life after death.

The gods brought fertility to fields, livestock, and people. Hathor, the goddess of love, brought men and women together, and the male god Min conferred sexual prowess—orgiastic festivals were held in his honor.[32] Goddesses like Isis helped fertility and motherhood, and women wore frog, locust, and monkey charms to enhance fertility. Married couples would pray to the gods for children, and some would visit the tombs of dead relatives, appealing to them to use their influence with the gods.[20] And, like some people today, they had statues of various gods in their homes.

Pharaohs

Pharaohs were more than kings.[20] They possessed the secrets of the heavens and performed special ceremonies to avert major threats, like the destruction of the sun. Their power was absolute; government and religion were inseparable.

Temples

The temples owned as much as one-third of the usable land, including villages, slaves, cattle, and dockyards. The temple of Amen alone owned eleven million donkeys and 90 percent of the ships.[32]

The high priest of the god Amen was also the director of the royal treasury and all public buildings. Foreign conquests by the armies brought wealth that was divided by the army and the temples, but little would flow to the pharaohs, leading to conflict with the priests. (And the temples were exempt from paying taxes.)

Priests

Religion and priests were important in everyday life. The priests were well compensated; family members usually inherited these elite positions of society. (In the powerful temples, the pharaohs would attempt to appoint their relatives to important positions.)

The Egyptians believed that the gods lived in the temples and the priests' role was to administer to them; if the priests neglected their duties, problems from the gods would arise.[33] The priests performed magic and served as oracles conveying the gods' answers to yes or no questions submitted by the people.[34]

Like the gods, there were male and female priests. Temples of gods usually had male priests; temples of goddesses, female priests.

While on duty, the priests had to bathe four times per day; shave their heads and bodies daily; lower their gazes; avoid laughing; abstain from sex; not eat fish, goats, and some birds; and not wear clothes made from living animals (no wool).[35]

Each morning, the priest would break the seal to the door or gate of the temple, light a torch, say prayers, burn incense, wash the gods' statues, place fresh clothing and jewels on the statues, and place offerings of food and drink (from the people) nearby.[35]

Only the priest could enter the sacred areas of the temple and approach the gods' statues. (In a few large temples, the people had access to separate outer rooms containing the gods' statutes.) The people's donations of food, flowers, and other valuables were left in the outer temple courts, where they prayed and sang hymns to the gods.

At the end of the day, the priest would back out of the temple, sweeping away his footprints, and seal the area. He would take the food and offerings left by the people home to his family or use them in barter.

The small minority of people who could afford it built tombs and had their bodies preserved. Their children would bring ritual offerings, but with short lifespans, they probably didn't know their grandparents or take care of their tombs. So taking care of tombs became a valuable (paid in advance) business for the temples, but it appears that they did a poor job providing this service as time went on.[35]

Akhenaten

Administering to thousands of gods allowed the priests to accumulate much wealth and power and high priests responsible for the major gods, like Amen, held a lot of power. Up and down the Nile, workers built hundreds of temples to the gods, and temple construction costs and foreign wars took a toll on the royal treasury. Over time, the pharaoh's power weakened as the priests' power strengthened, and since a priest's duty was to interpret the will of a god and the pharaoh's duty was to fulfill that will, the priests could help set the agenda for the pharaoh.[36, 37]

In 1353 BC, Amenhotep IV became pharaoh, succeeding his father, who had ruled for thirty-eight years. His wife was the "most beautiful woman in the world." She was Nefertiti; one of his children (Tutankhamen—King Tut, the boy king) would rule after him. Like his father, Amenhotep IV was concerned about the growing power of the priests and those of Amen in particular.[36, 37, 38]

Some historians believe that he was so concerned that he decided to abolish all the gods, thus eliminating the power of the priests. So in the ninth year of his reign, he changed his name to Akhenaten and declared that Aten, the sun god, was the only God, and that he, as pharaoh, was the only intermediary between Aten and the people. The people would henceforth pray to the pharaoh, and he in turn would communicate with Aten. The pharaoh, being the "king on Earth" and go-between to the one God, Aten, the "king of heaven," made Pharaoh Akhenaten "the son of God" on Earth.[36,37,38,39]

He ordered the removal of Amen and other gods' names from the temples, and turned two thousand years of Egyptian religious tradition on its head. Out of the chaos, disorganization, and lack of control caused by thousands of gods, he tried to create order with one God. The priests, of course, weren't happy, as they'd lost their power and source of wealth.

But Akhenaten died only eight years later, and, after a short interim period, his son Tutankhamen became pharaoh at age nine. Influenced, pressured, subverted, and controlled by powerful advisors and priests, the young pharaoh restored the worship of many gods and destroyed the temples of Aten.[36, 37] The one God rule of eight years could not overcome two thousand years of Egyptian multigod worship. (At the age of nineteen, Tutankhamen died from an infection as a result of a broken leg.)

Ironically, as time went on, the powerful Amen priests began to realize that they could benefit from the worship of one God: more wealth, more power, and control of the people. So gradually, Amen became the one main God.[36, 37]

Like today, religion was about influence, power, and wealth.

And that's how many gods became one God.

Chapter 4

Moses and the Bible Writers

Moses (1393 BC–1273 BC) was born in Egypt and raised in the pharaoh's household and was well aware of Egyptian customs, culture, beliefs, and the latest knowledge. Here are some things that Moses and the Bible writers observed, believed, and couldn't understand.

Some Things They Observed

Multiple Gods

They watched the people worshipping thousands of gods and goddesses, as was the case in the land of Israel at the time.

Power of the Priests and the Pharaoh

They witnessed the power and control that the priests and the pharaohs had over the people as well as the wealth that they gathered.

Confusion

They saw disorder and disunity as a result of the worship of thousands of gods.

Akhenaten's Experiment

They noted that the adoption of one God by Akhenaten in 1344 BC, curtailed the wealth and influence of the priests, and one set of beliefs for one God made it easier to organize and rule the people.

Potential to Unite the People of Israel

Perhaps they figured that a one God religion would be a potential way to unite the tribes and people in the land of Israel, allowing it to develop into a viable state free from foreign intervention from the likes of Egypt. (So eventually the belief in one God was adopted by Judaism and later by Christianity and Islam.)

Some Things That Moses and the Bible Writers Believed or Didn't Understand

- They believed that the earth was only six thousand years old (instead of 4.5 billion years old).
- They perceived that the earth was much smaller than its actual size, that it was flat, and that it floated on water. They couldn't figure out the size of the earth, the fact that it's round, or the extent of other continents. To them, the earth probably wasn't much bigger than the Middle East and the several hundred miles surrounding it.
- They thought that the sun and stars revolved around the earth.
- They believed that the sun and the moon could stand still.
- They thought that stars were twinkling lights and didn't understand the enormity of the universe.
- They believed the gods caused solar eclipses, lightning, thunder, rain, and flooding.
- They imagined that the pharaoh, by communicating with the gods, kept the sun from crashing.

- They didn't know about atoms, molecules, bacteria, fungi, viruses, and germs.
- They thought that evil spirits caused disease and illness and, therefore, the cures were through prayers, magic spells, and amulets.
- They didn't know the causes of and couldn't prevent the high infant and childhood mortality rates.
- They didn't know the causes or ways to minimize or prevent tuberculosis, malaria, smallpox, plague, and other contagious diseases.
- They believed that the heart did the thinking and couldn't figure out its relationship to blood circulation.
- They thought that semen formed a baby's white parts and menstrual blood formed the red parts.
- They didn't know that harmful "medicines" like animal dung could kill people.
- They didn't know that rotten teeth could lead to death.
- They didn't understand basic hygiene, sanitation, and sewerage.
- They didn't know how to develop more effective farming methods, tools, and harnesses that could have reduced malnutrition and famine.
- They didn't know how to develop simple building devices like pulleys and cranes.
- They didn't know how to build bridges.
- They believed that women were subservient to men.
- They thought that miracles were common occurrences, and, due to their lack of knowledge, lots of events and actions would seem miraculous.
- There was so little knowledge and so much that they didn't understand that they needed thousands of gods to account for all the unexplainable things.

So, with this narrow, limited background and primitive base of knowledge, Moses wrote the first five books of the Bible; the other writers followed and wrote the rest.

The Bible

The Bible is a collection of writings considered sacred by Judaism and Christianity. It became the first mass printed book and is the best-selling book of all time.

The Old Testament (Hebrew Bible) was written over a thousand years from 1445 BC to 450 BC by over forty different writers.[40, 41] It describes the creation of the world, God's relationship with humans, the stories of Abraham and Moses, God's commandments, the right ways to live life, and the history of the Israelites. The New Testament was written over fifty years from 50 AD to 100 AD by a dozen different writers and contains the story of Jesus.[42]

Critics claim that the Bible contains a great many errors, falsehoods, contradictions, and impossibilities, such as the following.[43, 44, 45, 46, 47, 48, 49]

The Bible:

- Says that the earth is only six thousand years old, that it is flat, and that the sun revolves around it.
- Endorses such practices as slavery, rape, incest, pedophilia, genocide, burning at the stake, racism, death to homosexuals, murder, human sacrifice, death for worshipping another god, etc.
- Says that the following are punishable by death: adultery, homosexuality, bestiality, cursing your parents, work on the Sabbath, sex with your daughter-in-law, etc.
- Describes God directly killing three million people; twenty million more were killed by Noah's flood.

- "Miracles" like Noah's flood, the plagues of Egypt, and Moses's parting of the Red Sea are all explainable events.
- Is obviously the word of men (not God), with passages such as:
    - "The husband is to rule over his wife."
    - "Wives are to be subject to their husbands even when the husband is disobedient to God."
    - "Man is the image and glory of God, while woman is the glory of man."
    - "Man was not created for woman, but woman for man."
- Only describes events in the Middle East (where the Bible writers lived; they had no experience outside of the area) and ignores the rest of the planet: Asia, Europe, the Americas, etc.

Searching the Internet will provide many more examples, and while many of the critics have a hostile, angry attitude, they raise legitimate points.

Most people that I know who go to church regularly and believe in God have never completely read the Bible. When they were young, they might have read small snippets at religious instruction, or they hear small pieces at Sunday services. They don't read it because they don't have time or don't find it very interesting. But, Christian Fundamentalists, Hasidic Jews, and others believe that the Bible is a historically accurate record of actual events.

It obviously cannot be the "word of a real God," because there are too many mistakes, atrocities, calamities, etc. It is the "word of the times" during which the Bible writers lived.

Aesop's fables were written in a similar period of time (620 BC–560 BC) and contain stories and lessons on how to live life. Perhaps the Bible should be called *The Bible Fables*.)

Today, the shocking thing isn't the material contained in the Bible (written by men of limited knowledge three thousand years ago), but rather that people today believe it to be literally true, which, of course, is rather incredible, . . . and unbelievable.

"It is between fifty and sixty years since I read the Apocalypse, and I then considered it merely the ravings of a maniac."

—Thomas Jefferson (1743–1826), third president

"The Bible tells us to be like God, and then on page after page it describes God as a mass murderer."

—Robert A. Wilson (1932–2007), author

"Most people are bothered by those passages (of the Bible) they do not understand, but the passages that bother me are those that I do understand."

—Mark Twain (1835–1910), author

"Kill a man, and you are an assassin. Kill millions of men, and you are a conqueror. Kill everyone, and you are a god."

—Jean Rostand (1894–1977), French philosopher

"No actual tyrant known to history has ever been guilty of one-hundredth of the crimes, massacres, and other atrocities attributed to God in the Bible."

—Steve Allen (1921–2000), comedian

"The Bible contains six admonishments to homosexuals and 362 admonishments to heterosexuals. That doesn't mean that God doesn't love heterosexuals. It's just that they need more supervision."

—Lynn Lavner, comedian

"The Bible is a collection of honorable but still primitive legends which are nevertheless pretty childish."

—Albert Einstein (1879–1955), physicist

"The God of the Old Testament is arguably the most unpleasant character in all fiction: jealous and proud of it; a petty, unjust, unforgiving control-freak; a vindictive, blood-thirsty ethnic cleanser; a misogynistic, homophobic, racist, infanticidal, genocidal, filicidal, pestilential, megalomaniacal, sadomasochistic, capriciously malevolent bully."

—Richard Dawkins (1941–), biologist

"Whenever we read the obscene stories, the voluptuous debaucheries, the cruel and tortuous executions, the unrelenting vindictiveness with which more than half the Bible is filled, it would be more consistent that we call it the word of a demon than the word of God."

—Thomas Paine (1737–1809), founding father

"I doubt any god who inflicts pain for his own pleasure."

—Dr. McCoy (2218–), *Star Trek*

"It will yet be the proud boast of women that they never contributed a line to the Bible."

—George W. Foote (1850–1915), publisher

"Read the Bible, because we need more atheists."

—Penn (1955–) & Teller (1948–), illusionists

Chapter 5

Visible Ignorance vs. Invisible Perfection

Gods That "Talked"

Today, God doesn't speak to people except those under the influence of drugs or alcohol, those who are mildly to severely delusional, those who have a bad dream or hallucination, those who want to believe so badly that their minds play tricks on them, or . . . those who are lying, phony preachers.

But in Moses's time, gods were speaking to priests and pharaohs all the time. The priests at the oracles would ask the gods questions and get yes or no responses. (At least the priests were clever enough to keep the answer options simple. Apparently, the gods didn't like multiple-choice questions, and essay responses were out of the question.)[34]

So when God "spoke" to Moses, it was a normal event for those times because gods were "speaking" to thousands of priests. (In 1976, psychologist Julian James (1920–1997), theorized that in ancient times, the brain/mind was not as fully developed as today, and when a person thought and asked a question in their mind, the person believed that the answer in their mind came from a "god." While the theory is controversial, it is still discussed in academic circles.[50] And maybe some good-hearted people even today misinterpret that other voice in their heads as the voice of God.)

The Male God

In the ancient world, there were as many goddesses as gods. But Akhenaten changed that by instituting one God—a male. The high priests followed that up later with one God—Amen, a male. If Moses had been a female, today we would have one Goddess (a female), and Jesus would probably have been female (Jesusa). Some of the actual history is a bit fuzzy, and some researchers think that Moses and Akhenaten were one and the same person; others say that Moses lived before Akhenaten and, therefore, Akhenaten adopted the belief in one God from Moses. Sigmund Freud thought that Moses was an Aten priest who had to flee Egypt after the death of Akhenaten.[38] The point is that even without Akhenaten and Moses, eventually multiple gods would have evolved into one God, if for no other reason than it gave religious/political leaders more influence and control over the people since for most of human history, government and religion have been closely intertwined.

Two-Thousand-Year-Old Knowledge

Knowledge, compared to today, was very limited (to nonexistent) a few thousand years ago. Few people were formally schooled, and the little knowledge there was had been passed down verbally from one generation to the next.

If we lived then, we too would have believed that the sun circled the earth, the earth was flat, and comets were bad omens. We would have thought that illness and disease were caused by evil spirits, and that Zeus and Thor caused lightning. We would have believed in and prayed to many gods.

But we don't live then . . . we live now.

Today, would you go to a doctor who only practiced two-thousand-year-old medicine, use hygiene, sanitation, and sewerage practices of two thousand years ago, or use two-thousand-year-old technology to build buildings, get from here to there, and conduct your daily life, etc.? Of course not.

So, let me see if I understand this. **A few thousand years ago, some men (not women) who were totally incapable of figuring out the visible physical world right in front of their noses were somehow capable of figuring out the invisible world of a God.**

Why do people today believe this? Do they lack mental courage? Do they lack curiosity? Do they refuse to use their brains? Are they just totally naïve?

Why do we believe what we believe?

Chapter 6

Gods We Don't Need Anymore

Things that were not understood hundreds or thousands of years ago are understood today, thanks to the ongoing development of knowledge.

As a result, there are many gods that we don't need anymore.

Gods We Don't Need	**Reason**
Sun Ra, the sun god Aten, the sun god Sol, the sun god Sunna, the sun goddess Apollo, the sun god	Today, we know that the sun is 4.5 billion years old and that it will continue to burn for another 5 billion years or more. It is one of 200 billion stars in the Milky Way galaxy. Three-quarters of it is hydrogen and the rest mostly helium. It is 93 million miles from Earth, and it takes eight minutes for its light to travel to us.
Moon Khonsu, the moon god Thoth, the moon god Ariadne, the moon goddess	We now know that the moon was formed from the collision of a Mars-sized planet into the earth more than 4 billion years ago, and its brightness is a reflection of the sun's light. We know that it orbits the earth and causes the tides. We can

Gods We Don't Need	Reason
Dianna, the moon goddess Luna, the moon goddess Cerridwen, the moon goddess	accurately predict when normal events, like lunar and solar eclipses and the passing of comets, will occur.
Earth Geb, god of the earth Gaia, goddess of the earth	The earth formed soon after the sun, and 70 percent of its surface is covered in water. Its iron core creates a magnetic field that protects us from harmful radiation from the sun. We know that it's about 25,000 miles around the equator.
Sea Poseidon, god of the sea Neptune, god of the sea	The oceans moderate Earth's climate. Winds blowing over the oceans create waves, and earthquakes under the oceans create tsunamis.
Storms Thor, god of thunder Tefnut, goddess of rain Zeus, god of thunder Jupiter, god of storms Tempestas, goddess of storms	Storms occur not only on Earth but on some other planets, like Jupiter. Today, predicting the weather is pretty accurate. And while I may not be able to tell you exactly how lightning works, I know that Zeus doesn't cause it.
Nile River Anuket, goddess of the Nile Hapi, goddess of the Nile	At 4,175 miles, the Nile is the longest (but not the only) river in the world. Starting in Lake Victoria, it travels through nine countries.

Gods We Don't Need	Reason
Harvest Demeter, goddess of the harvest Ceres, goddess of the harvest	Offerings and prayers to these goddesses protected the agricultural cycle. While we can thank Ceres for giving us the word "cereal," we now know that successful harvests are a result of good weather and agricultural practices.
Pleasure Voluptas, goddess of pleasure Bacchus, god of pleasure	While we would perhaps like to blame a night partying with our friends on these gods, our significant others probably won't buy it.
Love Aphrodite, goddess of love Hathor, goddess of love Eros, god of love Cupid, god of love Venus, goddess of love	We know that invoking the names of these gods in pursuit of romance is pointless.
Marriage Hera, goddess of marriage Hymen, god of marriage Freya, goddess of marriage	With divorce rates approaching 50 percent, perhaps we should start appealing to these gods again.

Gods We Don't Need	Reason
Fertility/Pregnancy/ Childbirth Bast, protector of pregnant women Heget, goddess of childbirth Taweret, goddess of fertility Brigid, protector of newborn babies Min, god of sexual prowess	Advancements in medical knowledge, which have led to improvements in fertility and substantial decreases in infant and mother mortality since Ancient Egyptian times, have made these gods obsolete.
Magic Isis, goddess of magic Serget, goddess of magic Trivia, goddess of magic Morgan, goddess of magic	Most people today do not believe in magic, even though some current religious beliefs and practices appear to be "magical." (But we do believe in Penn & Teller.)
Writing Seshat, goddess of writing Muses, goddesses of inspiration Apollo, god of poetry	We don't blame these gods for "writers block," but maybe we should.
Fame & Fortune Fama, goddess of fame Fortuna, goddess of fortune	Many people today would like to find a way to invoke these goddesses.

Gods We Don't Need	Reason
War Horus, god of war Neith, god of war Sekhmet, goddess of war Menhit, goddess of war Athena, goddess of war Mars, god of war Odin, god of war	Praying to gods of war would now be unacceptable. Besides, we know that men plan, initiate, and conduct destructive wars, not gods.
Freedom/Justice/ Wisdom/ Peace/ Security Libertas, goddess of freedom Justitia, goddess of justice Ma'at, goddess of justice Minerva, goddess of wisdom Pax, goddess of peace Securitas, god of security	Would that it be true, that all we had to do is invoke these gods to secure the object of their influence.

So, what don't we currently understand? What mysteries are left that require a god?

While we don't have a solution for cancer yet, do any of us believe that an evil god of carcinogens causes it? And because we have brilliant researchers working on it, we know that eventually a solution will be found.

There are other mysteries, like what's on the other side of a black hole, why socks get lost in the washer/dryer, and why my golf score doesn't come down, that I don't believe require the assistance of a god.

What's left? What Big Remaining Questions do we need God for? Our search brings us to these questions:

- Where did we (humans) come from? Who Made Us?
- Where did all the other stuff (the universe) come from? Who Made The Universe?

Part II
Proteins & Protons

Chapter 7

Proteins

Q: Who Made Us?

A: God Made Us.

—from *The Baltimore Catechism of the Roman Catholic Church*

Life

Life is basically a self-sustaining reaction of proteins and nucleic acids that allows an organism to grow and reproduce.[1,2] We know what's alive and what isn't—plants and animals are alive; rocks and soil aren't. But, in fact, rocks have microbes on them and inside them, and a teaspoon of soil contains a billion bacteria and millions of fungi.[3]

Kinds of Life

While there are ten million animal species and several hundred thousand plant species, the overwhelming majority of species are microscopic single-cell organisms, which, despite their small size, form half of the world's biomass.[4] Simple organisms have been the dominant form of life on Earth throughout its history and are the main form of life today.[5]

All life is pretty similar—the basic biochemical machinery of bacteria is similar to the chemistry of our cells.[1] The difference is that bacteria are single cells while our bodies contain 30 trillion cells and 100 trillion microbes.[6] Microbes are tiny, single-cell organisms like bacteria, fungi, and algae. The majority of microbes are harmless and perform vital functions like helping to digest food, absorbing salt and water, and regulating blood sugar. They could live without us; we would die without them.[1, 6]

Where Do You Find Life on Earth?

Life can survive and thrive in a wide range of conditions.

Microbes have been found:

- in steaming hot springs at Yellowstone National Park, where the water is above the boiling point;
- in acidic environments in Rio Tinto, Spain; around thermal vents under crushing water pressure 36,000 feet deep in the Mariana Trench in the Pacific Ocean;
- high in the atmosphere;
- in the driest deserts;
- in the Chernobyl nuclear meltdown area;
- underneath the ice in Antarctica; and
- inside rocks two thousand feet under the ocean floor.[7, 8]

Life can be found everywhere on Earth.

Life Up Close: Human Cells

Cells are the basic building blocks of life.

Human cells are very small—it takes 1,250 cells to equal one inch. (Bacteria are far smaller; it takes ten thousand of them to equal an inch.) One of our hands contains 2.5 billion cells, and every hour, 30,000 to 40,000 skin cells die and fall off our bodies.[9, 10] Each cell has its own job. Some carry oxygen, some defend against harmful bacteria, some transmit signals, some respond to light, and the rest do hundreds of different jobs. There are two hundred different specialized types of cells in the body, such as, heart cells, liver cells, and muscle cells.[10] Great quantities of identical cells organized together make up tissue (like muscle tissue). Many tissues organized together make up organs (like the heart, skin, and brain).

Cells reproduce and make new cells by dividing in two so that an organism can grow and replace old, dead, or damaged cells. Since every cell is the descendant of a single fertilized egg cell, each cell (except for egg, sperm, and blood cells) contains the same DNA (which tells the cell what to do). And, even though they are smaller than tiny, a cell is composed of many parts including the following.[10]

- Cell Membrane: The cell membrane is the protective covering around the cell, allowing oxygen and nutrients to enter while keeping toxins out.
- Cytoplasm: Cytoplasm is a watery gel-like fluid inside the cell and contains such things as proteins and nutrients.
- Nucleus: The nucleus is the cell's command center, which tells the cell to divide, stop dividing, take on nutrients, and when to die.
- Amino Acids: The human body uses twenty different amino acids; our bodies produce some, and others are supplied by the food we eat. The protein in food is broken down in the intestines into amino acids, which then enter our bloodstream and are eventually reassembled into new proteins. Amino acids are so small that it would take

twenty-five million of them to equal one inch. Proteins are chains of a few to thousands of amino acids.

- Proteins: Proteins help our bodies grow, stay healthy, and work properly. They do most of the work in the cell and give cells their structure, transport, and store nutrients, heal wounds, repair tissue, and help remove waste. There are over two million types of proteins in the body, and they're also very small, up to eight million to an inch. Genes control protein production.
- Enzymes: Enzymes are complex proteins containing a hundred to a thousand amino acids that speed up chemical reactions so that the cell can grow and reproduce properly. There are over two thousand types of enzymes in the body.
- DNA: DNA (deoxyribonucleic acid) contains information used in the development and functioning of all living organisms. It's a blueprint that tells the cell which proteins to make. DNA is shaped like a spiral staircase containing millions of steps, each of which consists of four chemical units called nucleotide bases—adenine, guanine, cytosine and thymine.[10,11] The sequence of these units determines which protein the cell makes. All cells within a person carry the identical entire set of DNA, but each cell type (kidney cell, skin cell, muscle cell, etc.) can access only the section of DNA needed for that cell type. Before a cell divides, RNA copies the DNA so that each of the resulting two new cells carries the same exact DNA.
- RNA: RNA (ribonucleic acid) takes the copied information from DNA and leaves the cell's nucleus, going into the cytoplasm, where the correct amino acids are assembled into the needed proteins according to the DNA instructions.

- Chromosomes: The DNA in the nucleus is packed tightly into structures called chromosomes. If stretched out, the strand of DNA in the nucleus would be six feet long but less than one millionth of an inch thick.[12] Humans have twenty-three pairs of identical chromosomes in the nucleus of each cell—one pair is inherited from the mother (egg cell) and the other from the father (sperm cell). Each chromosome contains a single long strand of DNA containing as many as two thousand genes.
- Genes: A gene is an individual section of DNA that carries instructions used to produce a specific protein for a specific job. Genes contain the information to build and maintain cells and pass genetic traits to offspring. A human has twenty to twenty-five thousand genes (consisting of three billion nucleotide bases) contained in each cell. This is known as the human genome.[12,13]

In general, the genome remains quite stable and consistent over an individual's lifetime. Reproduction of cells and the DNA they contain is generally quite accurate, but because the DNA in each cell contains billions of nucleotide bases, mistakes in copying can happen. When an error occurs in the copying process, it's called a mutation. Very minor variations in DNA are common and occur in most people, but some mutations could lead to cancer and other illnesses, and almost all diseases have a genetic component.[12, 13]

Some genes govern processes that run the body and others determine physical and behavioral traits. All humans have the same genes, but very small differences in DNA sequences make each person unique. Genes are the basic unit of heredity. Like chromosomes, genes come in pairs (one from the mother, the other from the father) and contain physical traits and personality traits passed on by parents.[12]

Simple traits—determined by one single gene—are highly predictable. A dominant gene will produce brown hair and a recessive gene red hair but only if the dominant gene is missing.

Many genes are involved in determining complex traits, such as height, blood sugar, and susceptibility to disease, making it difficult to predict which traits an offspring will have. A person may not resemble their parents but may show traits from their grandparents or great-grandparents, and while a trait can skip a few generations, if it does show up, it must have been present somewhere in the past in an ancestor.[12]

Researchers have identified many human diseases and individual traits that are inherited through genes. A few examples of inherited traits are: hair color, eye color, skin color, blood type, shape of nose and eyes, vision, dimples, curvature of the thumb, types of earlobes, right- and left-handedness, freckles, curly/straight hair, allergies, tooth shape, muscle structure, body type, and manner of walking.[12]

Researchers are constantly coming up with new discoveries, like the fact that one specific gene may give the Monarch butterfly the ability to migrate long distances, according to Marcus Kronforst of the University of Chicago. The alteration of a single human gene may result in suicidal thoughts, according to researchers at John Hopkins, and some genes may prevent disease.

How Did Life Begin on Earth?

From the formation of the first ocean, it took over 500 million years for chemical reactions to create simple organic compounds such as amino acids.[14, 15] The first living single cells (life) appeared 3.5 billion years ago, perhaps in a tidal pool or a deep ocean thermal vent. Five hundred million years later, cells that could perform photosynthesis evolved.[16, 17, 18] Most of life on Earth depends on photosynthesis; plants turn carbon dioxide, water, and sunlight into food (sugars), generating oxygen as a waste product giving us our livable atmosphere.

Around a billion years ago, the first multicellular plants like algae emerged. While evolution may at first seem difficult to understand, it takes place over long periods; it took 2.5 billion years to go from single cells to multicellular algae and another 400 million years for the first multicellular animals like sponges to evolve.[1, 14]

Around 500 million years ago, the first vertebrate animals like fish appeared. The buildup of oxygen gradually formed the ozone layer providing tiny organisms emerging from the oceans protection from the sun's radiation. The first animals left the ocean 450 million years ago. The dinosaurs rose 230 million years ago and ruled for 160 million years before being wiped out when a six-mile-wide asteroid crashed into the Yucatan in Mexico, generating tremendous dust clouds that blocked out the sun, killing the food source—plants. Most large animals perished and one-half of all animal species went extinct, but small mammals (our distant predecessors) survived.[1]

Two million years ago, the first human predecessors appeared. Over time, brain size increased, allowing increased social skills and language development. As language became more complex, ideas could be exchanged more readily and passed down through the generations; primitive knowledge slowly started to accumulate.

The dinosaurs ruled for 160 million years before going extinct; humans have only been somewhat civilized for ten thousand years. Before thinking that we'll have a longer reign, keep in mind that, so far, theirs is sixteen thousand times longer than ours. If the whole history of life on Earth were compressed into a twenty-four-hour day, then modern humans have only been around for a few seconds at the very end of the day.

***Some Key Dates in the History of Life on Earth*[19, 20]**

Event	How Long Ago
Earth formed	4.5 billion years ago
First ocean	4 billion years ago
First life, single cells (ancestor of all life today)	3.5 billion years ago
First photosynthesis (creating oxygen)	3 billion years ago
First multicellular life	2 billion years ago
Multicellular plants (like algae)	1 billion years ago
Multicellular animals (like sponges)	600 million years ago
First vertebrate animals (fish/amphibians)	500 million years ago
First land plants (like fungi)	475 million years ago
Insects	400 million years ago
Reptiles	300 million years ago
Dinosaurs	230 million years ago
Mammals	200 million years ago
Birds	150 million years ago
Flowers	130 million years ago
Dinosaur extinction/rise of mammals (half of all animal species go extinct)	70 million years ago
Primates	60 million years ago
Great apes	20 million years ago
Human predecessors	2 million years ago
Modern humans	200,000 years ago
Modern humans spread from Africa	150,000 years ago
Beginning of agriculture	10,000 years ago
First civilizations—Middle East/Egypt/India/China	5,000–6,000 years ago

At the Very Beginning

The earth possessed the right qualities for life to emerge—the right temperature range, liquid water, the right gravity, the right chemical elements, and an electromagnetic field (caused by its iron core) to protect life from harmful radiation from the sun.

In 1953, Stanley Miller, at the University of Chicago, simulated the conditions of a young, primitive Earth. He mixed steam, methane, ammonia and hydrogen; gave it an electrical charge; and a week later detected the formation of amino acids, a basic building block of life. His experiment showed that natural organic compounds could generate biological life compounds—amino acids.[1, 21]

Maybe life began as a very simple cycle—a natural chemical reaction that repeated itself—and along the way the little cycle acquired a cover of protective chemicals, giving birth to the first cell.[1, 14, 22] All current cells use the same basic chemical units and amino acids.[23] Today, we don't know exactly how this simple life started but does that mean that God started it? No. It means we haven't quite figured it out yet, just like we haven't quite figured out cancer, but we're getting there.[1, 14]

Inevitability

So, after the first simple cells, maybe next were amino acids, then proteins, then RNA, maybe a virus, then DNA over a period of many millions of years. And once DNA appeared, it might have completely replaced earlier life forms. Life may be something that naturally emerges as part of the chemistry of any planet possessing the right conditions.[1, 14, 18]

Evolution

Evolution is "gradual change." Cars evolve, airplanes evolve, fashion evolves, technology evolves, and plants and animals evolve—they all gradually change.

In biology, evolution is about changes made over many generations to the traits of large groups of the same plants or animals, adapting to a changing environment, giving those groups advantages or benefits.[24, 25]

Darwin

In 1831, Charles Darwin left England on a five-year research voyage on the ship *HMS Beagle*. Among its stops were the Galapagos Islands, a chain of thirteen major and over a hundred smaller islands located six hundred miles off the coast of Ecuador, South America. Because of the Galapagos' relative isolation, it is home to many unique animal species, including fourteen species of finches (birds). These birds descended from an ancestor from the mainland a few million years ago, and while nearly identical to those on the mainland, their beaks are different, having evolved over time based on the types of food available. Darwin observed that they had different size and shape beaks depending on whether each individual finch ate small seeds, large seeds, flowers, cactus flesh, leaves, or insects. He returned to England in 1836, continued his research, and in 1859 published his theory of evolution, *On the Origin of Species by Means of Natural Selection*.[24] ("Species" are groups of animals or plants that are similar and capable of reproducing with their own kind, like the "human species.")

At the time, in England, the prevailing view was that God had created the world in six or seven days and that the earth was six thousand years old. Darwin was the first to suggest that all life on

Earth descended from a single ancestor cell billions of years ago.[26, 27] His ideas were immediately attacked by the Church, but, in time, his theory of evolution became a major scientific discovery.

Causes of Change

Sometimes, scientists do a very poor job of explaining things in simple terms. While the biology of evolution can be complicated on the genetic level, the basic concept is simple. Large groups of plants or animals acquire beneficial changes slowly over time, improving their chances to survive and reproduce (and pass those beneficial traits on to their offspring).

Mutations and other changes to genes cause new traits to appear in offspring. Since mutations are random, they can be beneficial, harmful, or neutral (which happens most of the time). If it is harmful (which is rare), the offspring may not survive. If the mutation is beneficial, the offspring should do better in the world and reproduce, making the beneficial trait spread. An animal that gets a mutation allowing it to run faster is more likely to live and reproduce passing on the "faster" trait to its offspring. And eventually, as more mutations occur over long periods, new species may be formed.

Fifty million years ago, horses were small (the size of a fox) and lived in the woods, eating leaves, nuts, and fruits. Thirty million years ago, the earth became cooler and drier, forest was reduced in area, and grassy plains expanded. Over time, horses grew larger, ran faster to outrun predators on the open plain, and eventually evolved to the size they are today. Evolution of large organisms—animals—doesn't happen overnight; it took millions of years for the horse to grow to its current size.[28]

Darwin called the process of weeding out the harmful mutations and the spreading of beneficial ones *natural selection.*

Natural Selection

Darwin observed farmers breeding livestock and crops for the best traits and called this process *artificial selection.* The farmers were "selecting" the traits that they wanted in livestock (bigger, more meat) and crops (insect- and disease-resistant). In the natural world, he saw those traits that gave plants and animals better chances for surviving and reproducing as being "selected" in a sense by nature and called this natural selection. So mutations that cause beneficial traits to appear in offspring, while initially rare in a population, may eventually over time become common (because the ones without the trait die off and therefore do not reproduce). That's evolution.

Peter and Rosemary Grant have conducted current-day studies of Darwin's finches on the Galapagos Islands.[27] In 1977, a drought on the islands resulted in fewer small seeds and more large seeds. Because of fewer small seeds, many of the birds with smaller beaks starved to death and those with the bigger beaks survived. In 1978, they measured the surviving population and observed that the average beak size was larger than before the drought. Since many of the small-beaked birds had died and the surviving large-beaked birds had reproduced, passing on large beaks to their offspring, the overall population evolved to more large-beaked birds. Large beaks had been "selected" by nature. However, in 1984 and 1985, heavy rains created more small seeds and few of the large ones. More of the small-beaked birds survived and produced small-beaked offspring. Evolution had reversed, and the birds with the favorable trait (small beaks) survived and reproduced.

In England, before the Industrial Revolution, most of the peppered moths were speckled white in color and few were dark black. The speckled white blended in with lichen on trees, giving them camouflage from bird predators. As time went on, pollution turned the trees to a darker color, so the speckled white moths stood out and were eaten by the birds; the dark black moths blended in and survived. As a result, over time, the overall moth population's coloration changed from mostly speckled white to mostly dark black since the speckled white ones were more readily eaten and thus did not produce offspring. The white color was not a beneficial trait and was "evolved" out of the population. An individual moth didn't change colors or evolve—it just lived or died depending on its color. So white moths "evolved" to black moths because the white ones were eaten, leaving only the black moths to reproduce.[29]

Why Can't We Observe Evolution Happening?

Human evolution typically takes place over hundreds, thousands, and millions of years, which makes it difficult to observe changes taking place. It takes a long time for new traits to show up, and just as long to lose features that we don't need any more, like body hair, tonsils, the appendix, wisdom teeth, and the tailbone. A baby's strong grip is from millions of years ago, when babies' grasped and held onto their mothers' thick body hair. Our ancestors lived through times of famine, and a gene evolved that helped them store fat. Today, food is more abundant and this "thrifty gene," which helped long ago, may now be making people retain unhealthy amounts of fat, leading to obesity.[30]

Adult mammals cannot drink milk due to their inability to digest lactose, but in humans, when we domesticated cows several thousand years ago, a gene mutated and gave us an enzyme allowing us to digest milk. In the isolated town of Limone, Italy, a gene mutation occurred in 1644 resulting in a protein that quickly removed fats from arteries, thus reducing the risk of cardiovascular disease. This mutation and its benefits have been passed down through the centuries to today's descendants.

Scientists have recently discovered a gene found in 87 percent of the people of Tibet allows them to live more easily at high altitude. The gene was inherited by Tibetans forty thousand years ago from an ancient extinct group of humans. It keeps their blood from thickening in thin air, making them better suited for the high mountains.[31] Carlos Bustamante, a biologist at Cornell University, says that up to 9 percent of our genes are currently undergoing rapid evolution and these genes are primarily involved with the immune system, sexual reproduction, and sensory perception.

In humans, a generation is twenty years—the time from birth to producing offspring. But the timescale for bacteria is much shorter since it only takes twenty minutes to go from one generation to the next, allowing us to actually observe evolution taking place. In one year, a single bacteria produces 26,000 generations of bacteria. Humans need 525,000 years to produce that many generations. That's why we can observe bacteria evolving.

Bacteria became resistant to penicillin over a twenty-year span in which they produced over 500,000 generations (which would take over ten million years for humans to produce this many generations). The bacteria did not "evolve" a resistance to penicillin—a few had a resistance to penicillin—all the others were wiped out. So the few with the resistance kept reproducing, resulting in a population of bacteria resistant to penicillin. Other antibiotics were developed after penicillin, like methicillin and vancomycin, and were initially effective, but eventually bacteria "evolved" to become resistant to them.

Similar resistance has been observed with pesticides. Many insects reproduce several generations within weeks. While many of the insects will die with the first pesticide application, those that have the right resistant gene mutation will live and reproduce offspring carrying that immunity, making the pesticide ineffective.

Is Evolution a Fact or Just a Theory?

Evolution is a fact—it has been scientifically proven through research, investigation, the fossil record, and DNA analysis.[32, 33] Scientists sometimes use indirect words like "theory" to describe reality—it's just the way they talk. A much better word would have been "explanation."

So, think of the word *explanation* substituting for *theory* in

- the theory of evolution,
- the theory of special relativity,
- atomic theory,
- the theory of gravity, and
- the big bang theory, to mention a few "theories," all of which are facts.

So evolution is not subject to "belief." It's not okay for someone to say that he or she doesn't "believe" in evolution any more than a person can say that he or she believes that the earth is flat, 1 + 1 = 7, or the moon is made of Swiss cheese. Astoundingly, almost half of the people in the United States believe that God created humans directly, and evolution had nothing to do with it.

The fossil records prove the step-by-step changes that resulted in whales developing from their land-dwelling mammal ancestors over millions of years. And the fossil remains of countless other plants and animals demonstrates the different steps they took during their evolution.[32, 33]

Human genes match 98 percent to chimpanzee genes, 97 percent to orangutans, 90 percent to cats, 80 percent to cows, 75 percent to mice, 60 percent to fruit flies, 60 percent to chickens, 50 percent to bananas, 26 percent to yeast, and 18 percent to weeds. The amount of difference between our genes corresponds to how long ago in the past our genetic lines diverged. For example, we shared a common ancestor with chimpanzees five to seven million years ago and with orangutans fourteen million years ago. Would a god have made humans and bananas with 50 percent of the same genes? Of course not—they'd be totally different. The reason we share 50 percent with bananas is because everything—all plants and animals—descended from a single common cell ancestor billions of years ago.[34, 35]

What Does it All Mean?

Billions of years ago, basic elements and chemicals organized into reproducing molecules, and life began. That simple life form, through evolution over millions of years, has resulted in all living things that exist today as well as those that have gone extinct (99 percent of all life), like the dinosaurs.[1, 36]

Change results in evolution. Evolution doesn't stop—it's a part of life. We and all living things are constantly evolving. Evolution is continuous.

Coming Soon

People have been breeding animals and plants for thousands of years. A male and female that exhibit desirable traits are bred together, and the offspring that show the desirable traits are mated together again, and so on over many generations. Cattle grow larger and produce more milk, chickens produce larger eggs, wheat becomes more insect- and drought-resistant, and dogs exhibit desired traits.

So humans throughout history have been selecting traits and "forcing" evolution. Now comes genetic engineering, which eliminates the breeding process by directly manipulating genes in bacteria, insects, plants, fish, and mammals to produce desired results. Genes are modified to create biofuels and pharmaceutical drugs. Crop resistance to pests is improved while crop yields are increased. Animals have genes manipulated to increase animal health and for use in medical research. Fish have been modified to grow faster.[37]

In the future, gene therapy could be used to cure human diseases like Parkinson's, cancer, diabetes, heart disease, schizophrenia, and Alzheimer's. Parents could select traits desired in babies. And soon synthetic biology may result in new forms of life that will generate caution and controversy.

Life in the Universe

Ninety-five percent of life on Earth is based on six elements: carbon, hydrogen, nitrogen, oxygen, phosphorus, and sulfur.[1, 38] These elements are spread everywhere throughout the universe. In fact, the complex organic molecules necessary for life may have been formed in the dust and debris particles circling the sun before the earth was even formed—life on Earth may have been seeded by these particles.[39]

On Earth, life can be found under the crushing pressure of the deepest oceans, buried beneath thick Antarctic ice and in the harshest deserts. It's hard to find a place without life. Life is very resourceful—life is tenacious.[40]

Life may exist (or have existed) on Mars, in the atmosphere of Venus, on comets, and in the subsurface oceans of Jupiter's moons

(Europa, Ganymede, and Callisto), and Saturn's moons (Titan, Enceladus, and Mimas).[40, 41]

The surface of Europa is frozen twelve miles thick because it's so far from the sun. But underneath this ice, the enormous gravity of Jupiter stretches and compresses Europa, generating heat and keeping its subsurface oceans liquid. These oceans hold twice the volume of water of the earth's oceans. Geysers continuously shoot hundreds of tons of water miles into space. Bill Nye the Science Guy suggests that we fly a robotic craft through the geysers and analyze the water for signs of life.[42]

In the Milky Way galaxy alone, there may be forty billion earthlike planets located in life-friendly zones.[43] Scientists like Carl Sagan and Stephen Hawking said it would be unlikely for life not to exist other places.[44, 45] Harvard astronomer Avi Loeb, thinks that it's possible that shortly after the big bang, life was everywhere in the young, expanding universe.[46] (Someday in the future, when intelligent life is discovered on another planet, do you think that Jesus would have visited them as well? Or, would God have sent a different Son to visit them, since being God, He could have an infinite number of Sons, but apparently no Daughters).

Q: Who Made Us?

Correct A: Very early on in Earth's 4.5-billion-year history, life began slowly, from carbon, hydrogen, nitrogen, oxygen, phosphorous, and sulfur, combined with the right temperature range, gravity, liquid water, protection from the sun's radiation, and perhaps an electrical charge. It developed into single cells, which are the ancestors of all life on Earth today. Over billions of years, it evolved into all life existing today, including us.

—From Facts, Science, Knowledge, Truth

"Those who would legislate against the teaching of evolution should also legislate against gravity, electricity, and the unreasonable velocity of light and also should introduce a clause to prevent the use of the telescope, the microscope, and the spectroscope, or any other instrument of precision which may in the future be invented, constructed, or used in the discovery of truth."

—Luther Burbank (1894–1926), botanist

"Evolution is one of the best documented, most compelling, and exciting concepts in all of science."

—Stephen J. Gould (1941–2002), biologist

"Basically we are chimpanzees with about two percent more intelligence and a little less hair."

—Ted Turner (1938–), businessperson

Chapter 8

Protons

Okay, so who made all the elements—the building blocks of life—and the universe itself?

Q: Who Made The World?

A: God Made The World And Is The Creator Of Heaven And Earth And All Things.

—*Baltimore Catechism*

The Big Bang

Key Discoveries

Some remarkable discoveries and advances in knowledge happened between 1900 and 1930, giving us a better understanding of the universe. Albert Einstein's theory of special relativity (E = MC2) found that space and time are interwoven into what is called space-time, and his theory of general relativity found that massive objects could distort space-time. Ernest Rutherford and Niels Bohr enhanced our understanding of the atom. Edwin Hubble discovered that the Milky Way is only one of many galaxies.

Werner Heisenberg and Erwin Schrodinger explained quantum mechanics—the physics of the very small (atoms and smaller). Georges Lemaitre discovered that the universe is expanding and theorized that since it was expanding, at some point in its past it was a single point, and from that single point, it expanded in the big bang.

Protons, Neutrons, and Electrons

Everything is made of atoms. Atoms are made of protons, neutrons, and electrons. Protons and neutrons are made of quarks. Quarks, electrons, and other subatomic elementary particles are the building blocks of all matter.

These particles are incredibly small. A cell in our body can contain 100 trillion atoms.[47] A drop of water can contain 5 sextillion (five with twenty-one zeroes) atoms.[48] The nucleus of an atom (which contains protons and neutrons) is 100,000 times smaller than the atom itself.[49] Quantum mechanics is all about the action and behavior of these subatomic particles.

Creation of the Universe

Over 13.5 billion years ago, the universe consisted of a single point of infinite density and a temperature of 1,000 trillion degrees, which in a fraction of a second went through rapid expansion called inflation. The big bang was not an explosion of matter into empty space—it created space and time, expanding at a rate faster than the speed of light.[50, 51] (This expansion continues today at a much slower pace.)

As the early universe expanded and cooled, elementary particles of matter formed, and 380,000 years after the big bang, nuclear reactions created hydrogen and helium and then all the other elements and matter. The first star formed 100 million years after the big bang and the first galaxies 200 million to 500 million years later.[52, 53]

What Caused the Big Bang?

We don't exactly know what caused the big bang. We don't even know if it had a "cause." We do know that on the quantum level, subatomic particles like electrons, photons, and positrons can go in and out of existence randomly.[52, 53, 54]

Perhaps the universe started in a similar manner. Perhaps it resulted from an enormous black hole. Because we don't know how it came to be and because the math and physics are incredibly complicated and difficult to understand doesn't mean that a God started it. It means that we haven't figured it out yet. Stephen Hawking, Leonard Mlodinow, Victor Stenger, Robert Matthews, Laurence Krauss, and other physicists say that the big bang can be explained by the laws of physics—that the universe could have created itself from nothing—and that God is not necessary to explain its existence.[54]

The Spread and Acceptance of Knowledge

Some five hundred years ago, most people believed that the earth was stationary and the sun revolved around it. Then came Copernicus and Galileo, who described how the earth revolved around the sun. Today, we accept that to be true, although we may find it extremely difficult to describe the mathematics and physics associated with the sun's gravitational force that keeps the earth in orbit. But without needing to grasp the detailed mathematical understanding of how this happens, we "know" that it's true.

I don't exactly "know" how electricity works (flow of electrons, etc.) but I do know that lightning is not caused by gods like Zeus and Thor. I don't know how to build an airplane or produce a car, but that doesn't mean that I believe God gives information to people in order to do so.

An Associated Press/GFK survey of March 2014 found that virtually all people believe that smoking causes cancer. That was not the case in the 1950s when smoking was far more popular. In 1964, the surgeon general released the first report about the dangers of smoking, followed for decades by constant media attention, so that now, people "know" that smoking causes cancer.

The survey found that almost all people accept the fact that genes help determine who we are. That was probably not widely known just a few decades ago—lots of information and press coverage has educated the people.

While Darwin and the theory of evolution removed the need for a God to explain the existence of humans some 150 years ago, the survey found that only 55 percent of the people accept evolution. Interestingly, the survey found that 85 percent of the people believe that the overuse of antibiotics caused the development of drug-resistant bacteria. Since the bacteria became drug resistant through the process of evolution, why would there be a difference in the answers? Perhaps because people have read about and heard about the drug-resistant bacteria on TV repeatedly. But human evolution is not generally discussed. It's not newsworthy enough.

The big bang theory is almost a hundred years old. In that time, scientists have researched and discovered evidence that supports and proves the theory. As a result, 99.9 percent of the members of the National Academy of Sciences accept it as fact.[55] The AP/GFK survey found that about 50 percent of the population does not believe in the big bang. Perhaps this is due to the big bang being the most complicated concept of all the survey questions, coupled with the fact that it does not get as much press coverage.

So knowledge grows, spreads, and is accepted over time (even if people don't quite understand it). The speed at which it is accepted depends on how much information, discussion, publicity, education, and press coverage occurs.

In less than a hundred years, through the development of science and knowledge, we've gone from little understanding of the universe to the point where we can say that our understanding of our dynamic, expanding universe, while not complete, allows us to be much closer to explaining its existence without the need of a God.

Moses and the Bible writers were at a tremendous disadvantage. They had no knowledge of atoms, protons, neutrons, electrons, neutrinos, quarks, bosons, molecules, cells, bacteria, algae, fungi, viruses, cytoplasm, cell nuclei, amino acids, proteins, enzymes, DNA, RNA, chromosomes, genes, the elementary building blocks of life, the age of the earth, the earth's magnetic field, gravity, evolution, the size of the earth, the revolution of the earth around the sun, the planets, the galaxies, the size of the universe, electromagnetism, the strong force, the weak force, the theory of relativity, quantum mechanics, the big bang, and millions of other things that we know and take for granted today. (The average person today possesses an incredible amount of knowledge compared to Moses. If we could make Moses magically, instantly time-travel to today, knowing that he'd possess the least amount of knowledge of virtually anybody in this country, would you believe his stories?) So it was entirely logical for them to believe that a god caused everything in their world. If we lived then, we would have believed the same.

Q: Who Made The World?

Correct A: We Don't Exactly Know, But Knowledge Is Building And We Believe That The Universe's Existence Can Be Explained Without The Need Of A God.

—*From Facts, Science, Knowledge, Truth*

So if we don't need a God for humans to exist and, if we don't need a God for the universe to exist, why do we believe? Moses had an excuse—a total lack of knowledge. What's our excuse?

In time, virtually all people will believe in evolution and the big bang.

Then what happens to God?

"Nature is capable of building complex structures by processes of self-organization; simplicity begets complexity."

—Victor Stenger (1935–2014), physicist

"It is now becoming clear that everything can—and probably did—come from nothing."

—Robert Matthews (1959–), physicist

"The laws of gravity and quantum theory allow universes to appear spontaneously from nothing. Spontaneous creation is the reason there is something rather than nothing, why the universe exists, why we exist."

—Stephen Hawking (1942–), physicist

Part III
Time & Distance

Chapter 9

Our Perception of the World

Attempting to comprehend the vastness of the universe is the key to understanding our insignificance.

Our Shrinking World

Ancients

Imagine you lived thousands of years ago. You farmed, hunted, or fished and lived in a small settlement or village. You probably didn't travel very far, and when you did, it was on foot. Your "world" was what you could see with your eyes. You didn't know what existed beyond your village or region. Folklore may have told you that the very ancient peoples came to your area from far, distant places. You were born, lived, and died within a few miles, with virtually no understanding of the rest of the world, which was mysterious and incredibly big. Knowledge was slow to develop.

Columbus

The world grew a little smaller in 1492 when Columbus set sail to find a westerly route to the Orient and instead crashed into

America. His ships traveled at an average speed of two to five miles per hour and took ten weeks to sail from Europe to the New World, with only a magnetic compass and an hourglass as navigation tools.

My Grandparents

All four of my grandparents left the poverty and discrimination in Ireland for opportunity in America around 1895. They traveled in steerage on steamships like *The Majestic* and *The British Prince* that took eight days (fifteen to twenty miles per hour) to cross the Atlantic. There were no planes crisscrossing the ocean.

The night before they left Ireland, their friends and family gathered and held a party called an American wake because they knew they would never see each other again. Even so, their "world" was smaller than Columbus's.

Today

Today, we routinely drive fifty to seventy-five miles per hour. Could the ancients or Columbus have comprehended this speed? Maybe they had a sense of it from viewing horses, deer, dogs, or birds. We jet over to Ireland in about five hours, search for our roots, and return home anticipating our next return. Could they have comprehended flying at five hundred miles per hour over the Atlantic in a matter of hours from Europe to America?

Perceptions change over time. A few hundred years ago in Europe, churches were the biggest buildings (along with castles). Their height let large windows bring magically colored light (from stained glass) inside to the faithful who attended a few times each week. Their imposing size reinforced the Church's power and influence (as castles did to peasants).[1] Today, if you were asked to name a large building, you probably wouldn't say a church since in many cities they have been dwarfed by much larger buildings.

We can understand bigger buildings, larger distances, and faster speeds than the ancients, Columbus, and our grandparents. How will our offspring perceive time and distance in a thousand years? In a hundred thousand years? A million years?

Our Importance

As we travel in our daily world, we see trees, houses, buildings, and other people, and our perception is that we must be important in this "world." And we are—as is our dog.

But does this everyday feeling of importance lead us (and our ancestors, like Moses) to feel that we are of major importance to the universe? Does this perception distort what a God would be like and look like?

When you look down from an airplane, humans are virtually invisible, and when an astronaut looks down from the moon, whole cities disappear. If we could comprehend the size of the universe, would that change our perception of our own importance—how small we would be to the eyes of a God? It was logical for Moses and the people of his time to think that they were rather important, as they believed that the earth was only as big as the Middle East (and maybe parts of Africa and Asia) and was at the center of the universe, the center of creation. And, if those things were true, we today would be very significant in the whole scheme of things.

But just as we can comprehend faster speeds and bigger distances than our ancestors, unlike them, we also understand the vastness of the universe and our almost total insignificance.

How Big Is a Million (1,000,000)?

We can easily grasp "thousands," like $1,000, that it's 1,500 miles from Boston to Florida, and that it costs $30,000 for a car and $200,000 for a house.

But a "million"?

In 1996, at the one hundredth running of the Boston Marathon, officials estimated that a million spectators lined the route, which is hard to comprehend. Since the University of Michigan football stadium holds 100,000 people, the marathon spectators would have filled up ten football stadiums, which I can begin to grasp. I also get (but maybe not understand) million-dollar salaries for athletes and entertainers or million-dollar homes.

Breaking large things into smaller units makes comprehension easier.

How About a Billion (1,000,000,000)?

The Big Dig in Boston cost $15 billion—a tremendous amount of money—and other new construction projects cost many billions of dollars.

In my yard, I have ten majestic oak trees. Estimates are that each tree holds two hundred thousand leaves—that's two million leaves to rake in my yard each year. There are over a thousand trees in my whole neighborhood or two billion leaves (2,000,000,000) each year.

A Trillion (1,000,000,000,000)?

The 2013 Federal Budget[2] was about $4 trillion. Hard to imagine, but if you break it down into major components it helps.

2013 US Federal Budget

Medicare	$940 billion
Social Security	$882 billion
Military	$672 billion
Interest on Debt	$246 billion

Agriculture	$154 billion
Veterans Benefits	$140 billion
Treasury	$110 billion
Labor Department	$102 billion
Transportation	$99 billion
Education	$72 billion
International	$60 billion
Homeland Security	$55 billion
Intelligence	$53 billion
Housing	$46 billion
All Other	$369 billion
	$4 trillion

And if you break each of these categories (like Transportation) down into smaller pieces, $4 trillion comes into focus.

Unlike Moses and our ancestors, we now use numbers in the millions, and billions, and trillions, and by breaking them down into understandable pieces, we begin to comprehend them.

So we've got a handle on a trillion dollars, but can we comprehend a trillion miles? A quadrillion or quintillion or sextillion miles?

Chapter 10

Our Neighborhood in the Universe

The Solar System

The solar system formed 4.5 billion years ago from a gigantic dust and gas cloud. Over millions of years, this dust and gas began to clump together and gravity, pressure, centrifugal force, and magnetism made the clumps of debris bigger and bigger. Most of the mass (material) concentrated at the center of this swirling vortex, creating the sun. Farther from the sun, similar forces gradually created the planets that revolve around the sun.[3] Four smaller, rocky, metallic planets—Mercury, Venus, Earth, and Mars—are closest to the sun. The farther away planets are Jupiter and Saturn (made mostly of hydrogen and helium, like the sun) and Uranus and Neptune (composed of ice made from water, ammonia, and methane).[4]

Jupiter, the largest planet, is over three hundred times the size of Earth. There are also six dwarf planets like Pluto, 173 moons, 3,298 comets, and hundreds of thousands of asteroids (the left over debris that hasn't yet clumped onto a planet).[5]

Our Star: The Sun

Our star is an ordinary star, made of 75 percent hydrogen, almost 25 percent helium, and various other metals. Through nuclear processes, it turns 400 million tons of hydrogen into helium each second. Since its birth 4.5 billion years ago, it has used up one-half of its initial supply of hydrogen; its surface temperature is 10,000°F.[6]

And while it's located 93 million miles from Earth, it takes only eight minutes for its light to reach Earth. The sun's power drives Earth's climate and weather, and through photosynthesis, plants absorb its light, generating oxygen for our atmosphere. The plants also provide food, and through their decay over millennia, energy (oil, gas, and coal).

In 2012, a solar flare from the most powerful storm on the sun in 150 years narrowly missed Earth. If we had been in its direct path, it would have knocked out electrical power and damaged electrical equipment that would have taken days and months to repair and get back to normal.[7]

In about a billion years, the slowly increasing temperature of the sun will evaporate all the water on Earth. In five billion years, it will enlarge to over 200 times its current size, engulfing Earth, and will then cool over billions of years.[6]

Our Planet: Earth

Earth began forming about 30 million years after the solar system started forming, and it took 10 million to 20 million years for it to gravitationally pull together enough debris to get to a decent size. Soon after it formed, it was impacted by a smaller planet, tilting Earth's axis, and throwing debris into orbit that gradually coalesced into the moon.[3] For 500 million years, it swept around its orbit of the sun, collecting debris, being impacted by meteors and asteroids, and building to its current size. Earth measures 25,000 miles around; the sun is 2,715,000 miles around. It would take 1.3 million Earths to fill up the sun.

Asteroids and comets brought water, and as the planet cooled, clouds formed and rain developed, creating the oceans. While Earth was forming, heavy elements like iron sank to its center, giving it a metallic core and creating a magnetic field around it. This magnetic field protects us from some of the sun's harmful radiation, prevents our atmosphere from being stripped away, and allows life to develop. Venus and Mars lack a magnetic field, causing what's left of their atmospheres to slowly disappear into space.

Small asteroids—fifteen feet across—strike Earth once a year. Larger ones (twenty-five feet across) hit Earth every five years with the energy of an atomic bomb, which is mostly absorbed in the atmosphere. One-hundred-and-fifty-foot asteroids hit every thousand years with an explosive force of a thousand atom bombs, and every hundred million years, a dinosaur-killer asteroid six miles wide hits.[3]

Chapter 11

Big and Beyond Big

Big: Our Galaxy, the Milky Way

When I was young, I'd look up and marvel at all the twinkling stars in the universe. I didn't know that the stars twinkled because of Earth's atmosphere and that I wasn't looking at the universe, but only the Milky Way galaxy, which is massive in its own right.[8] All the stars we see with our naked eye are part of the Milky Way.

It's called the Milky Way because it appears to be a dim, glowing "milky" band across the night sky. In 1610, Galileo, through his telescope, discovered that the "band of light" was actually individual stars.[9]

The Milky Way is a gigantic star- and planet-making factory, converting hydrogen into heavier elements like iron and carbon from which planets and humans and everything else is made. Stars are formed in regions known as molecular clouds and spend their lives fueled by the nuclear fusion of hydrogen into helium, radiating energy into space. They're not spread uniformly across the universe but are grouped into galaxies that can contain ten million to a hundred trillion stars.[8] The nearest visible star to our sun is Alpha Centauri, which is 25 trillion miles away.

The spiral-shaped Milky Way contains 200 billion stars (and ten times that number of planets) and measures 600 quadrillion miles across.[10] It also contains gas, dust, black holes, dark energy, and dark matter.

Our sun and solar system orbit the center of the Milky Way galaxy. While it takes a year for Earth to orbit the sun, the Milky Way is so vast that it takes the sun, traveling at five hundred thousand miles per hour, 250 million years to make one orbit of the center of the galaxy.[9]

Scientists are now discovering other "Earthlike" planets in the galaxy. The first was discovered in 1992 and now hundreds have been found.

And galaxies do collide, but due to the enormous distances between stars, they usually pass through each other without destruction. They can also merge to form a large single galaxy; the Milky Way is currently in the process of absorbing two smaller galaxies. The Milky Way and Andromeda galaxies are moving toward each other at 250,000 miles per hour and will collide in four billion years.[11]

Beyond Big: Our Universe

The Beginning

The universe was caused by the big bang over 13.5 billion years ago, and the same basic elements are spread throughout the universe. All stars and planets are made of these elements—there are no marshmallow fluff planets or peanut butter moons. The same physical laws that regulate our solar system are consistent throughout the universe. Water (liquid and frozen) and life organisms are probably abundant throughout the universe.[12]

Vastness

The universe is enormous (200 billion galaxies) and is constantly expanding. Gravity binds together each galaxy even as all the galaxies are flying farther apart. It's so vast that to try to imagine its scale is impossible.[12]

But let's try.

How Many Stars Are There?

Assume that there are 200 billion galaxies and, on average, each galaxy holds 200 billion stars, and that each of these stars is represented by a grain of sand. A University of Hawaii student project counted over three million grains of sand in a one-cup container. That means that the average backyard inground swimming pool will hold 800 billion grains of sand—a tremendous amount.

But to account for all the stars in the universe, we would need fifty billion swimming pools!

Or, if we filled up railway tanker cars with sand, we would need a train 285 million miles long that would encircle the earth eleven thousand times.

The world's largest oil tanker ship (which if stood on end would be taller than the Empire State Building) carries eighty-four million gallons and can fill the gas tanks of five million cars. We would need nine million super tanker ships to hold all the stars.

And the number of planets would be ten times the number of stars!

Traveling

The manned Apollo spacecraft traveled at a speed of twenty-four thousand miles per hour.

If we were able to build a spaceship that could travel at a million miles per hour, here's how long it would take to get around the universe.

Destination	**Distance from Earth**[13]	**Travel Time at 1 Million Miles Per Hour**
The moon	240,000 miles	15 minutes
Mars	48,000,000 miles (48 million)	48 hours
The sun	93,000,000 miles (93 million)	4 days
Jupiter	400,000,000 miles (400 million)	17 days
Saturn	800,000,000 miles (800 million)	34 days
Pluto	3,500,000,000 miles (3.5 billion miles)	146 days
Alpha Centauri (the nearest star)	25 trillion miles (25,000,000,000,000)	2,800 years
Sirius (the brightest star)	50 trillion miles (50,000,000,000,000)	5,600 years
The Big Dipper	806 trillion miles (806,000,000,000,000)	90,000 years
The next galaxy, Andromeda	15 quintillion miles (15,000,000,000,000,000,000)	1.7 billion years (1,700,000,000 years)
Far edge of the universe	267 sextillion miles (267,000,000,000,000,000,000,000)	30 trillion years (30,000,000,000,000 years)

These distances are truly astounding. Traveling at a million miles per hour, it would take 2,800 years to get to the nearest star (Alpha Centauri); and, almost two billion years to get to the next galaxy (Andromeda)!

Intelligent Life

If the universe is awash with water and basic life, has intelligent life evolved on other planets and why have we never been contacted?[14]

- The distances are just too enormously large.
- We have only been capable of detecting a distant civilization's signal for a very brief period of time—fifty years. Maybe these signals weaken and dissipate over great distances, maybe they're not focused in our direction, or maybe they're sending a very different kind of signal that we aren't yet capable of recognizing.
- A signal traveling at the speed of light from a distant intelligent civilization would take hundreds, thousands, or millions of years to get to us, and after sending the signal, the civilization may have gone extinct. A message traveling at the speed of light (186,000 miles per second or 671 million miles per hour) is no match for the vastness of the universe. That message if traveling to us from the area of the Big Dipper would take 134 years to get here. This means that the starlight that you see tonight looking up at the Big Dipper started on its journey to us 134 years ago. If it were traveling to us from the next galaxy—Andromeda—it would take 2.5 million years to get here!

Other Universes: The Multiverse

Years ago, "the universe" meant everything, all that existed. But maybe it's just an enormous collection of galaxies. Some scientists theorize that there may be many "universes," each existing in separate space/time (like separate soap bubbles) that don't interact with each other. These are also called parallel universes or alternative universes and, theoretically, there may be an infinite number of them.[15]

What Does This All Mean?

The universe is so enormous that we are totally incapable of grasping its size. It would take a spaceship traveling at a million miles per hour, almost two billion years to travel to the next galaxy (Andromeda), never mind the travel time to the other 200 billion galaxies.

So an ant on Earth is a billion trillion times more significant to Earth than a human is to the vastness of the universe. Our existence is so trivial, so inconsequential, that it is massively dwarfed by the enormity of our insignificance.

This comprehension, this realization, this acceptance, is the key to understanding that:

- There very likely is no God.
- If there is a God, it's not as we imagine (and as has been described by Moses and our ancestors: an old guy with a beard and long flowing robes).
- This God cannot possibly know (or care) that we exist. (Do you think that God cares about the smallest, tiniest bacteria on Earth? Aren't we humans far smaller than that bacteria compared to the enormity of the universe?)

To think that if there is a God, He, She, or It knows that we exist doesn't make sense. And to think that this God talks to some preachers and individuals is beyond all reason. These preachers and individuals are deluded, lying, or hoping so hard that their imaginations play tricks on them.

So it is the height of human ignorance, it is the height of human conceit, to think that if there is a God, He cares about you. And the concept of Moses's God is not believable.

Over time, as knowledge develops, gods disappear. Today, we no longer need a god to explain our or the universe's existence, so this being the case, what kind of (not needed) God/religion do we have now?

Part IV
Our Current God/Religion

Chapter 12

The "Ancient Egyptian" Catholic Church

When it comes to god and religion, we want to believe that ours is special and unique. It's not. It's built on the concepts and practices of previous religions. In the case of Christianity, most of its ideas, concepts, rituals, and practices have been taken from the religion of ancient Egypt.

Christianity Has Taken These Concepts and Practices from the Ancient Egyptian Religion[1, 2, 3, 4, 5, 6, 7, 8]

Afterlife: Heaven/Hell, judgment of the dead

Altar

Amen: In Egypt, it was the god Amen; now it's said at the end of prayers

Baptism

Chalice

Charms: Religious jewelry

Cross: In Egypt, it was the Ankh

Devils/Demons/Evil Spirits

Flagellation

Flowers on gravesites

Halos: In Egypt, it was the sun disk

Holy Water: In Egypt, it was the "tears of Isis"

Hymns

Immortality of the soul

Incense

Minor gods in Egypt: Now called saints in Christianity

Mitre

Monstrance

Mother of God: In Egypt, it was Isis; now it's Mary

No Taxes Paid: By the temples and now the Church

Offerings

One God

Prayer Beads

Prayers

Priests

Religious candles

Religious processionals: Through the streets

Resurrection: In Egypt, it was Osiris; now it's Jesus

Ritualized burial

Rituals

Sanctuary

Scapular: In Egypt, it was the amulet

Son of God: In Egypt, it was Horus; now it's Jesus

Staff

Statues

Popes' and bishops' dress/clothing: Very similar in appearance to the pharaohs

The Trinity: In Egypt, one trinity was Isis, Osiris, Horus; another was Amen, Ra, Ptah; now it's the Father, Son, and Holy Spirit

Transubstantiation/the Eucharist: In Egypt, it was the flesh of Osiris; now it's the flesh of Jesus

Wall drawings, statues: In homes

Wine: In religious ceremonies

(Some believe that the Egyptian god Horus was the model for Jesus, and Horus and his mother Isis were the models for the Madonna and child.)

Later on, Christianity tried to extinguish paganism by taking over their holidays and building churches on the sites of pagan shrines. Here's a few of the holidays that were taken.

Easter

Easter was taken from the pagan festival of Eostre, the goddess of the dawn. This celebration was held at the spring equinox, when early crops began to sprout, signifying new life, rebirth, and the end of winter.[9] Its symbols were the rabbit for fertility and the egg for new life.

Christmas

Jesus's birthdate is unknown. This was the pagan celebration of the winter solstice, a day when the sun reached its lowest point in the heavens and would begin its rebirth, signaling the hopeful coming of spring.[10] (Some important pagan gods had birthdays on December 25.)

Halloween

This was a Celtic harvest celebration during which candles were lit for the souls of the dead. It marked summer's end and the beginning of winter. Eating, drinking, and games were part of the celebration.[11]

May Day

This was a pagan celebration of fertility and flowers and in old times, May first was considered the first day of summer (the summer solstice of June 21 was midsummer).[12] The Church turned it into a celebration of the Virgin Mary wearing a crown of flowers.

The Church took over these holidays, and others, forcing these beliefs on the people. In time, this oppression will weaken and fall, and these holidays will be restored to their original meanings.

Not Unique

So Christianity is not a unique religion—it stole, copied, and mimicked older ideas.

Chapter 13

The "Holy" Roman Catholic Church

Q: How Will You Know The True Church?

A: The True Church Of Jesus Christ Is The One, Holy, Catholic, Apostolic Roman Church.

—*The Baltimore Catechism*

Definitions

Catholic = Universal

Apostolic = Relating to the Apostles

Holy = Good, virtuous, sinless, pure, highly moral

So, how "holy" has the Church been throughout its existence?

Perhaps inspired by the murder, mayhem, and genocide conducted by God in the Old Testament, the Church has its own litany of "unholy" actions, including persecution, murder, and genocide.

Pagans

We remember the Hollywood movies that depicted pagan Romans torturing and feeding Christians to the lions in the

Colosseum. The roles reversed in 380 AD when Christianity was declared the official religion of the Roman Empire, and the pagans who refused to be baptized lost their income and property and were declared void of all rights. Pagan religious services were punishable by death, pagan temples were destroyed, pagan priests were killed, and pagan libraries were burned.[13] Estimates are that hundreds of thousands of pagans were killed.

Heretics

In the first few centuries AD, the central power of the papacy was limited. Christian communities believed what they wanted, decided how to "follow Jesus," and developed their own worship practices. Around 180 AD, Irenaeus (a saint), the bishop of Lyon, France, observed different Christian sects and beliefs forming and decided that there should be a central authority and uniform set of beliefs. Those who didn't comply were called "heretics" and would need to be reeducated or punished.

Laws against heresy began appearing in 380 AD in the Roman Empire. Augustine (a saint; 354 AD–430 AD) said, "Error has no rights" and was in favor of fines, imprisonment, banishment, and flogging to convert heretics. As time went on, laws were passed banning heretics from holding public office, working in certain professions, and forfeiting their civil rights. In 1085, Pope Gregory VII (a saint), said "the killing of heretics is not murder."[13, 14] (Prior to 1100 AD, the persecution of heretics seldom led to executions.)

The first Inquisition was started in 1184 and was run by the Dominican clergy in association with local bishops. Pope Gregory IX made the Inquisition a permanent institution in 1229, and in the beginning, it only was concerned with Christian heretics. It was heresy to not agree to the authority of the local bishop or the pope, read a Bible (only the clergy could do that), quote the Bible, translate the Bible into local languages, eat meat on Fridays, criticize the clergy, refuse to pay Church taxes, or violate any other Church doctrine or rule. Punishments included fines, imprisonment, torture, property confiscation, and death. People under investigation had to wear two pieces of red cloth in the form of an X as a public display. In serious cases, those who confessed were strangled and their bodies burned. Those who didn't confess were burned alive.[14, 15, 16]

The Cathars were a community of Christians spread throughout southern France. They were a peaceful group that believed in nonviolence but did not recognize the pope or pay Church taxes. The Waldensians were a Christian group in southern France and northern Italy that gave money to the poor and preached the gospel. They denied the authority of priests and the pope, and rejected the Mass, indulgences, confession, penance, Latin, saints, and killing. They allowed women to preach and didn't pay Church taxes. There were many other similar Christian groups, such as the Paulikians, Runcarians, and Josephites. They were all excommunicated (although they never belonged to the Roman Catholic Church) and were declared heretics.

Then, Pope Innocent III's Crusader army, which numbered up to 500,000 troops, invaded Beziers, France, on July 22, 1209. The pope's Crusaders captured a hundred "heretic" prisoners, gouged out their eyes, cut off their noses and lips, and sent them back into the town led by a prisoner with only one remaining eye in order to terrorize the inhabitants. Thousands were mutilated, dragged behind horses, crucified, stoned, impaled, and in other ways slaughtered—men, women, children.[13,14,15,16] The town was burned. Next, fell the town of Carcassonne, then Mirret, then Marmande; others followed.

Thomas Aquinas (a saint; 1225–1274), thought that heretics should be "eliminated from the world by death." By 1255, the Inquisition had spread throughout central and western Europe (except in England and Scandinavia), and it went on for seven hundred years. The Spanish branch of the Inquisition began in 1478 with the backing of King Ferdinand and Queen Isabella. The infamous Inquisitor—Torquemada—targeted Muslims and Jews who had converted to Christianity but who were suspected of practicing their old religions. In 1492, all Jews who had not converted were expelled from Spain, and those who remained were subject to the Inquisition. Many went to Portugal, many went to Turkey, many died. The Inquisition spread to Spain's colonies, including North America, Central America, and South America. The last execution in Spain was in 1826.

The Portuguese Inquisition started in 1536. Many of the Jews who had left Spain were targeted. It spread through Portugal's colonies, including Brazil. It ended in Portugal in 1821. In-depth descriptions of diabolical tortures, depravities, and slaughters can be found in many other books.

No one knows how many were killed and slaughtered. Estimates range from tens of thousands to hundreds of thousands.[17] Maybe the best, most accurate records are still sealed in the Vatican secret archives. In 1908 the Vatican's office of "Inquisition" was renamed "The Sacred Congregation of the Holy Office," and in 1965 the name was changed again to "The Congregation for the Doctrine of the Faith."

The Church was the Taliban of the day—it conducted horrors like those that occur in the Middle East today.

Witches

In 1484, Pope Innocent VIII ruled that cats were unholy creatures and should be burned along with the witches who owned them. The Church believed that women were inherently evil and sexual and in league with the devil. During the era of witch hunting (1480–1800), many witches were hung, crushed by stone, drowned, or burned at the stake. Women were tortured to confess that they flew through the night and had sex with the devil. Sometimes entire villages were exterminated. When the European climate cooled during the Little Ice Age from 1600 to 1800, Pope Innocent VIII blamed witches for the freezing weather, failing crops, rising crime, and famine.[18]

In the 1600s, the Puritans in Massachusetts created a totalitarian church state in which violation of church rules could result in flogging, pillorying, hanging, cutting off ears, and boring through the tongue with a hot iron. Twenty witches were killed, and 150 were imprisoned.

From 1480 to 1800, estimates are that from thirty thousand to a few hundred thousand "witches" in total were killed.[13]

Jews

When Christianity became the official religion of the Roman Empire, marriage between a Christian and a Jew became punishable by death, and converting to Judaism became a criminal offense. Jewish property was seized, and synagogues were randomly destroyed. Cyril (a saint) expelled Jews from Alexandria and from many other countries.

To warm up for the First Crusade, the French Crusaders crossed into Germany in 1096, where they massacred thousands of Jews in the cities of Speyer, Worms, Mainz, and Cologne.[19] Some historians call this the First Holocaust, beginning a thousand-year program of organized attacks on the Jews of Europe." [13, 20, 21] (Supposedly, some bishops offered the Jews protection in exchange for money.)

In 1298, 140 Jewish communities in Austria and Bavaria were destroyed and thousands were killed. In 1321, Jews were accused of inciting criminals to poison wells, and five thousand were burned alive at the stake. In 1347, the Black Death, which had originated in the Far East, reached Europe. Over five years, twenty-five million died in Europe. Rumors were that the Jews, helped by Satan, spread the plague—thousands were executed.[13, 20]

And it continued through the centuries, culminating in the Holocaust. So, the Church set a precedent for Hitler, showing that persecution and killing of Jews could be done with success and be accepted by the general population.[16]

The Crusades

The Crusades were a series of wars against Muslims, pagans, Orthodox Christians, Mongols, Cathars, Hussities, Jews, and other political enemies of the popes. The popes claimed that they were working on behalf of Christ. The several hundred thousand who became Crusaders took a public vow and were forgiven by the pope for the future sins they were to commit in the Crusades.

Between 1095 and 1291, there were seven major Crusades and numerous minor ones. The clergy instituted two hundred years of violence, all in the name of God.[13, 21]

First Crusade 1095–1099

Pope Urban II launched the First Crusade in 1095. Its goal was to take control of the Holy Land from the Muslims, and its battle cry was "God Wills It."

In 1097, the Crusader army attacked the Turks at the city of Antioch, Syria, near the Turkish border. Prior to entering the city, they catapulted the heads of previously captured Muslims into the city to spread fear. Upon taking and pillaging the city, they massacred a hundred thousand Turks.

In August 1098, low on food and water, the city resorted to cannibalism. Radulph of Caen wrote, "Our troops boiled pagan adults in cooking pots, and they impaled children on spits and devoured them grilled." In 1099, they entered Jerusalem and killed almost every inhabitant, thereby "purifying" the city. In some places they waded ankle-deep in blood; more than sixty thousand were killed. But Jerusalem was captured—the Crusade was a success.[22, 23, 24, 25]

Second Crusade 1147–1149

The Muslims took over some territory near Jerusalem, which led Pope Eugene III to call for a new Crusade that failed to win any major victories in the Holy Land but was successful in driving the Muslims from Lisbon, Portugal. At the time, Bernard Clairvoux (a saint) said, "Cursed be he who does not stain his sword with blood."[26]

Third Crusade 1189–1197

In 1187, Saladin, the Sultan of Egypt and Syria, united the Muslims and took control of Jerusalem. The pope proclaimed that the Muslim capture of Jerusalem was punishment for the sins of Christians in Europe.

Richard I of England (Richard the Lionheart) recaptured some of the cities near Jerusalem but not Jerusalem itself. On August 20, 1191, he had 2,700 Muslim prisoners decapitated in view of Saladin's army; in response, Saladin killed all the Christian prisoners he had captured.[22,23] On September 2, 1192, they signed a peace treaty giving the Muslims control of Jerusalem while allowing Christian pilgrims and merchants to visit. Richard's decision not to attack Jerusalem would lead to the Fourth Crusade.

Fourth Crusade 1202–1204

The Fourth Crusade had originally intended to sail to Egypt and work its way up the coast to Jerusalem. But short on funds, the leaders were persuaded by the son of the deposed ruler of Constantinople to go there and retake the Eastern Orthodox city. Constantinople was the largest, most cultured city in Europe at the time. It was a stronghold of Christianity, and due to its strategic location, was a barrier for Europe against the advances of Islam.

The Crusaders captured Constantinople and burned a large part of the city, leaving a hundred thousand homeless. They raped the Eastern Orthodox nuns, murdered the priests, and plundered the Hagia Sophia, the greatest church in Christianity at the time, taking the Church's treasures, silver statues, and holy books.[24, 25]

This ended the chances of reuniting the Roman Catholic Church and the Eastern Orthodox Church. Pope Innocent III spoke against the Crusaders but accepted the Church's share of the stolen wealth from Constantinople.

Constantinople was recaptured by the Greeks in 1261, but in 1453, the Turks took it over, renaming it Istanbul. In 2011, eight hundred years later, Pope John Paul II apologized for the slaughter to the Greek Orthodox Church.

Fifth Crusade 1217–1221

Pope Innocent III and Pope Honorius III called a Crusade to retake Jerusalem in 1217. The Crusaders were defeated and surrendered to Sultan Al-Kamil, and both parties agreed to an eight-year peace treaty.

Sixth Crusade 1228–1229

This was another attempt to retake Jerusalem. Emperor Frederick II had repeatedly vowed to conduct a crusade but was slow to act; for dragging his heels, Pope Gregory IX excommunicated him in 1228. Eventually, Frederick embarkd with a small force and with very little fighting was able to arrange a ten-year treaty with Sultan Al-Kamil of Egypt, allowing Frederick to rule over Jerusalem as Al-Kamil had other problems needing his attention.

Seventh Crusade 1248–1254

In 1245, the son of Al-Kamil retook Jerusalem, and the Crusader army was defeated.

For all the Crusades, the total killed was estimated at several million people.[27]

Religious Wars

In addition to the Crusades, there were religious wars in Europe between Catholics and Protestants.

French Wars of Religion

Conflict and killing between Catholics and Protestants in France began before the 1562 start of the French Wars of Religion and continued after the 1598 end. Eight separate wars were fought during this thirty-six-year period. Nobility, soldiers, men, women, and children were assassinated, massacred, pillaged, mutilated, castrated, drowned, hanged, and burned. It's estimated that two million to four million died—there is no count of the maimed or wounded. Conflicts and killings continued sporadically for more than a hundred years, and eventually, Catholicism became the official state religion of France.[27, 28]

Thirty Years' War

In 1618, the Protestant nobility opposed the newly crowned Emperor Ferdinand of the Holy Roman Empire as they feared Ferdinand's Jesuit training and desire to spread Catholicism. Two of Ferdinand's representatives were unceremoniously seized and thrown out of a castle window (they survived), sparking conflict and war. The war that started over religious issues evolved to political ones and engulfed most of Europe for thirty years, with the fighting done primarily in today's Germany.

At the time, armies were self-funded and raised money by looting and extorting; entire areas were devastated. War, famine, and diseases like typhus, dysentery, and plague crushed the population. The population in the German states was reduced by 20 percent to 40 percent, and half of the males were killed. The Swedish Army alone destroyed 2,000 castles, 18,000 villages, and 1,500 towns in Germany. In total, between three million and twelve million people died.[27, 29]

The Peace of Westphalia in 1648 resulted in the Holy Roman Empire being broken up into numerous independent states. The map was rearranged, and political power began to rule over religious power.

Native Peoples

The Americas

Starting in 1492, Columbus brought disease, violence, and Christianity to an estimated 100 million natives in the Americas. Everywhere he landed, he planted a cross requiring obedience to the Church and claiming the land for Spain.

While he, and all who came after him, may not have understood the spread of smallpox and other diseases that they brought with them (that killed two-thirds of the native population), they did understand the violence they inflicted on the remaining population that all but wiped them out in the name of Christianity.[13, 16, 30]

Slavery

In 1452, Pope Nicholas V placed the Church's stamp of approval on the slave trade, and between 1502 and 1853, an estimated twelve million Africans were brought to the Americas.[31] In 1993, Pope John Paul II made a public apology for the Church's involvement.

Stolen Generations

From 1910 to 1970, Aboriginal children in Australia were forcibly removed from their families. Up to one-third of all Aboriginal children were taken, often suffering physical and sexual abuse, and nonpayment for their work. In all, a hundred thousand children were taken to orphanages, church missions, and foster homes.[32] In 2001, Pope John Paul II issued an apology. He also apologized for the behavior of Catholic missionaries in China in colonial times.

The List Goes On

Between 1922 and 1996, ten thousand young girls in Ireland were sent to the Magdalene laundries to labor unpaid under the supervision of nuns. The harsh, prisonlike conditions and slave labor stole their youth, dignity, and self-worth. In 2014, the Irish government issued an apology.[33]

Catholic clergy were involved in the massacre of several hundred thousand people in Rwanda in 1994.[16] Etc.

"Holy"?

These are only some of the "highlights" of Church actions; probably hundreds or thousands of other brutalities occurred. So, it's quite clear that the last adjective that we can use to describe this Church is "holy."

Chapter 14

Some Criminal Popes (& Infallibility)

As previously mentioned, since the Vatican is an independent state and wraps its historical records and archives in secrecy, discovery of the nature and amount of crimes committed by popes through the years is difficult.

Celibacy was optional up to 1139, when it became required. The pope's infallibility concerning doctrines of faith and morals, long thought to be true, was formally confirmed in 1870, and applies not only to popes from 1870 on, but also to all popes throughout the entire history of the Church.[34]

Some "Infallible" Popes[35, 36, 37, 38, 39, 40]

Pope Damasus I (366 AD–384 AD)

Following the death of Pope Liberius, Damasus competed with Ursicinus to be the next pope. Seeking safety, Ursicinus and his supporters sought refuge in a basilica, and in the surprise massacre that followed, Damasus and his followers killed 125 to 150 supporters of Ursicinus. None of Damasus's followers was killed; he is now known as St. Damasus.

Pope Stephen III (768–772)

During the contest to become pope, Stephen's followers captured the other candidate, Constantine and his followers, blinded them, and ripped out their tongues.

Pope Pascal I (817–824)

Pascal was suspected of involvement in the blinding and beheading of his own followers in order to cover up a potential revolt against the King of Italy. Upon his death, he was denied burial in St. Peter's Basilica but is now known as St. Pascal.

Pope Boniface VI (896)

He was defrocked twice for immorality as a subdeacon and priest and died after fifteen days in office as pope.

Pope Stephen VI (896–897)

He had his predecessor, Pope Formosus, dug up, dressed in papal robes, and set on the papal throne to be interrogated for his "crimes." Not surprisingly, the corpse was found guilty and had its three fingers (the ones used in blessings) cut off. The body was stripped of the robes and reburied, only to be dug up again, and thrown in the Tiber River. His actions caused riots; he was deposed, imprisoned, and strangled.

Pope Sergius III (904–911)

He was rumored to have ordered the murder of his predecessor, Pope Leo V. Described as disgraceful and ruthless, he fathered an illegitimate son, who later became Pope John XI.

Pope John XII (955–964)

He turned the papal palace into a brothel and forced married women, widows and virgins, who had come to Rome as pilgrims to view the tombs of the Apostles, into sexual orgies. He had sex with his sisters and niece and raped the nuns. He ordained a ten-year-old boy in exchange for sexual favors, enjoyed invoking demons, and tortured, castrated, and murdered a cardinal. He was killed while in bed with one of his mistresses—by her husband.

Pope Boniface VII (984–985)

He claimed the papacy after having his predecessor Pope Benedict VI strangled. He robbed the Vatican treasury and fled to Constantinople. Benedict VII was made pope, and when he died, John XIV became Pope. Boniface returned to Rome and had Pope John XIV murdered. He died suddenly in 985, and his body was dragged through the streets.

Pope Benedict IX (1032–1045)

St. Peter Damian called him "a demon from Hell in the disguise of a priest." Pope Victor III said he was "so vile, so foul, so execrable." He filled the papal palace with orgies and rapes of men, women, children, and animals. Murder was common. Finally, he sold the papacy to his godfather (who became Pope Gregory VI) and married his cousin.

Pope Urban II (1088–1099)

He launched a bloody "holy" war against the Muslims known as the First Crusade. In order to attract recruits to fight, he forgave their sins in advance, thus beginning 200 years of war resulting in an incredible loss of life. He is now known as St. Urban.

Pope Innocent III (1198–1216)

He launched the Third and Fourth Crusades resulting in much killing, the rape of Constantinople, and a permanent break with the Eastern Orthodox Church. In 1209, he sent the Crusaders to Southern France to battle heretics. When soldiers asked their papal advisor how to tell the faithful from the infidel, the advisor supposedly said, "Kill them all. God will know his own." Twenty thousand were slaughtered.

Pope Gregory IX (1227–1241)

He permanently established the Inquisition in 1229 and brought the persecution and killing of Jews to new heights.

Pope Boniface VIII (1294–1303)

He was a power-hungry pope who clashed with many rulers, resulting in a great many dead. Dante placed him in the eighth level of Hell in *Inferno.* He had children with two of his nieces and was also accused of assassinations.

Pope Clement V (1305–1314)

He crushed the organization of Knights Templars (who previously were used to protect pilgrims traveling to the Holy Land), seized their properties and monies, and burned many of them alive at the stake.

Pope Urban VI (1378–1389)

Arrogant and angry, he tortured cardinals and set off fifty years of bloody wars over who was the real pope.

Pope Gregory XII (1406–1415)

When the rector of the university in Prague, Jan Hus, spoke out against abuses and corruption of the clergy and papacy, he was invited to appear before the pope to defend his views. His safety was guaranteed by the pope. Upon appearing, he was excommunicated and burned alive at the stake. The pope reportedly said that "when dealing with heretics, one is not obligated to keep their word."

Pope John XXIII (1410–1415)

The first Pope John XXIII became pope one day after becoming a priest and was accused of piracy, murder, incest, and selling indulgences.

Pope Eugene IV (1431–1447)

He tortured and killed followers of his predecessor Pope Martin V and endorsed slavery as long as the enslaved weren't Christians. He condemned Joan of Arc of witchcraft (who was later made a saint by Benedict XV in 1920).

Pope Paul II (1464–1471)

His opponents were tortured and killed, and he extorted enormous wealth. He required Jews to wear identifying yellow handkerchiefs (a tactic later employed by Hitler) and forced them to run naked in the street for his amusement.

Pope Sixtus IV (1471–1484)

He established the Spanish Inquisition, which cruelly tortured and killed many, and was a lover of boys, many whom he eventually nominated to be cardinals.

Pope Innocent VIII (1484–1492)

He supported the torture and killing of witches and magicians and had many illegitimate children. He was a master of bribery, selling papal offices to raise money, and sold pardons for various sins to support his vices.

Pope Alexander VI (1492–1503)

He had several mistresses and fathered numerous illegitimate children. Bribery and greed were commonplace, and orgies were frequent. He used his daughter, Lucretia (Borgia), to seduce wealthy merchants and marry her off to gain vast dowries. Then, as pope, he would have the marriage annulled and repeat. He was poisoned to death.

Pope Julius II (1503–1513)

Elected pope by bribery, he had several illegitimate children and was constantly scheming. Nicknamed "The Warrior Pope."

Pope Leo X (1513–1521)

He raised money by selling forgiveness for sins including murder, theft, and incest.

Pope Paul III (1534–1549)

He slaughtered heretics in Germany and Italy, had several illegitimate children, and was a proponent of slavery. He poisoned his mother and niece and had a bishop and two cardinals killed over a theological dispute.

Pope Julius III (1550–1555)

He devoted his life to personal pleasure. He adopted a teenage beggar found in the streets and shared his bed and looted the papal treasury to renovate his mansion.

Pope Pius XII (1939–1958)

Hitler: Due to its network of parishes, the Church was well informed continuously of the ongoing Holocaust. Yet, the pope did not speak against the Nazis and the atrocities.[41,42,43] In 1941, Harold Tittman of the US Delegation to the Vatican, urged the pope to speak out and condemn the atrocities. He received a response that the Vatican wanted to stay neutral. (One reason given for staying neutral was that the Church wanted to maintain its "historic" position as a peacemaker, which is unbelievable nonsense—the Church's "historic" position for 1,500 years was as a warmaker!) Supposedly, once US officials told the pope in late 1942 that the Allies would win the war, the pope advised the Bishops in Germany and Hungary that it would be to their political advantage to speak out against the Holocaust.[44,45,46]

Croatia: In 1929, Ante Pavelic formed a radical group—Ustase—in Croatia seeking Croatian independence from Yugoslavia. When Germany took over Yugoslavia in 1941, Croatia became a German puppet state, and Pavelic became an ally of Hitler.

Pope Pius XII was in favor or Croatian independence as he saw Croatia as "an outpost of Christianity." He met with Pavelic. Pavelic and the Ustase were fiercely Catholic and proceeded to persecute and kill Serbs, Jews, Gypsies, and dissidents with the help of local Catholic clergy. As many as four hundred thousand were killed in massacres and death camps in 1941. The Pope was well informed of the slaughter and the Catholic clergy involvement; he said nothing and did nothing.[47] A British Foreign Office memo of the time reportedly called the pope "the greatest moral coward of our age."[48]

Baptized Jewish Children: In 2005, the Italian newspaper *Corriere Della Sera* disclosed the contents of a Vatican directive dated October 23, 1946, which had been recently discovered in a Church archive near Paris. It instructed that Jewish children who had been baptized during the war (in an effort to shield them from the Nazis) must not be returned to their parents, since the children were now Christians.[49] (If Jewish children weren't baptized, they could be returned to their parents or family.)

Ratlines: Ratlines were the escape routes used by Nazis, Ustase, and other genocidal criminals fleeing Europe after the war to South America and other countries. Some of these routes went through the Vatican, where the criminals were given rudimentary documentation that could be used to obtain forged passports.[50, 51]

Gold: Rumor has it that gold seized by the Ustase and the Nazis flowed to the Vatican Bank, where it may have been then transferred to South America or other banks. The Vatican refuses to let investigators review the records.

Pope Pius XII's Journey to Sainthood: In the Roman Catholic Church, recognition as a saint requires four steps.

Step 1: Initial investigation. If approved, the prospect for sainthood is called a "Servant of God."

Step 2: Proof of heroic virtues. If approved, the prospect is called "Venerable."

Step 3: Beatification. If the prospect has one miracle to their credit, they are called "Blessed."

Step 4: Canonization. If the prospect is credited with a second miracle, they are called "Saint."

In 1965, Pope Paul VI began the process of sainthood investigation for Pope Pius XII. In 1990, Pope John Paul II declared him a "Servant of God." In 2009, Pope Benedict XVI made Pope Pius XII "Venerable," which means he had demonstrated heroic virtues. In January 2014, Pope Francis said he would open the Vatican secret archives to scholars in order to evaluate Pope Pius's role in World War II; it was estimated that this process would take a year.

Father Peter Gumpel, a Jesuit historian and proponent of Pope Pius XII's sainthood, claims that there are several miracles attributable to Pope Pius XII including "one quite extraordinary." But in May 2014, Pope Francis said that the sainthood process for Pope Pius XII had stalled because "no miracle had been found."[52]

Are the researchers still combing through the Vatican secret files?

Pope John Paul II

Recently, Pope John Paul II was sainted. Critics say he was slow to acknowledge and respond to the child sexual abuse crimes of the clergy and only did so when public pressure demanded it. He gave sanctuary to Cardinal Law of Boston, who covered up and moved sexual predatory priests from parish to parish, allowing them to sexually abuse more and more innocent children. He promoted Hans Groer, a sexual predator, to archbishop and cardinal in Vienna. He backed Marcial Maciel of Mexico, founder of the Legion of Christ, who was a known, longtime sexual predator who had fathered illegitimate children and was a drug addict and embezzler but also provided a steady stream of money to the Vatican.[53, 54]

Cover-Up

Some may feel that the murders, treachery, depravity, etc. of the previous popes, while certainly repulsive, may, in the whole scheme of things, be not so bad for a two-thousand-year-old institution. (Is this like saying that Hitler did some good for Germany and Stalin for Russia?) The problem is that the crimes and corruption of these popes may only be the tip of the iceberg. It is said that during World War II, the Allies and average German citizens were unaware of the true extent of Hitler's extermination camps. Only at the end of the war, when the Allies liberated the camps, was the depravity, atrocities, evil, and death toll uncovered. Is this the same with the Church? The fall of the Roman Empire, combined with various invasions and wars and the rise in the power and control of the Church made the Church one of the chief keepers of European historical records for over 1,500 years. And the Vatican closely guards and keeps secret the historical records of these and all other popes. Do we need to demand that the Vatican's Secret Archives be opened to researchers to get a full picture of the Church's history of horrors?

Infallibility Impossibility

Based on just the previously described popes, the Church's doctrine of papal infallibility is, of course, preposterous—it is, of course,. . . impossible.

Chapter 15

The Institute for the Works of Religion

The Church throughout its history has been one of the all-time leaders in obfuscation—making things unclear, confusing, obscure, or evasive—and usually hiding behind words and pronouncements in Latin. The Institute for the Works of Religion is in reality the Vatican Bank, known as the most secretive bank in the world. Founded in 1942, it doesn't make loans but does receive deposits, makes money transfers, and manages $11 billion in assets. The bank's history has been defined by scandal, secrecy, and noncompliance with standard banking and antifraud measures.[55, 56, 57]

Michele Sindona became a friend and financial advisor to Pope Paul VI, who reigned from 1963 to 1978, and became quite close to Archbishop Paul Marcinkus, head of the Vatican Bank from 1971 to 1989. Huge amounts of money moved from Sindona's Italian banks through the Vatican Bank to Swiss banks. He was known in banking circles as "the shark" and also managed the Gambino crime families' heroin profits.

In 1972, Sindona purchased controlling interest in Long Island's Franklin National Bank; on October 8, 1974, Franklin

National was declared insolvent due to mismanagement and fraud, and the Vatican Bank lost $33 million in Franklin's failure.[56]

Pope Paul VI died in 1978.

Pope John Paul I took over. Sindona's friendship with Archbishop Paul Marcinkus (nicknamed "the Gorilla") caused Pope John Paul I to complain several times to the archbishop. On September 28, 1978, after just thirty-three days in office, Pope John Paul I was dead. The Vatican said the cause was a heart attack; no autopsy was performed.[58]

On July 11, 1979, Giorgio Ambrosoli, the lawyer put in charge of liquidating Sindona's Italian banks, was murdered after discovering massive fraud on the part of Sindona. (On March 20, 1979, Mino Pecorelli, a journalist who had been investigating Archbishop Marcinkus and the Vatican Bank and its ties to organized crime, was murdered.) On March 27, 1984, Sindona was found guilty of Ambrosoli's murder and sentenced to life in prison in Italy. While working on his appeal, Sindona decided to disclose everything, but on March 18, 1986, he was poisoned by cyanide in his coffee.

In 1982, Archbishop Paul Marcinkus was accused of being an accessory in the $3.5 billion collapse of Banco Ambrosiano (which was accused of laundering drug money for the Mafia).[57] Roberto Calvi, Chairman of Banco Ambrosiano, known as "God's Banker" because of his close connection to the Vatican Bank, was convicted of violating Italian currency laws and fled to London on a false passport. A few days later, on June 17, 1982, he was found murdered and hanging from Blackfriars Bridge.[56]

The Vatican Bank denied legal responsibility for Banco Ambrosino's failure but acknowledged "moral involvement" and paid $240 million to creditors.[59] Archbishop Paul Marcinkus avoided arrest by staying in Vatican City until the arrest warrant was dismissed in 1991 and returned to his home in Arizona, where he died in 2006. He was suspected of being involved in several mysterious deaths, including that of Pope John Paul I.[60]

In 2012, JP Morgan Bank closed the bank accounts it held for the Vatican Bank because the Vatican Bank wanted JP Morgan to wire transfer monies around the world without the required documentation of who the money was going to. Supposedly, JP Morgan told the Vatican Bank that it had to answer to the regulators; the Vatican Bank said that "it answered to God."[61] In 2013, Deutsche Bank closed down the Vatican's eighty cash machines for similar concerns.[57]

Currently the Vatican Bank's former general director, Paolo Cipriani, and his deputy, Massimo Tulli, are facing money laundering charges in Italian courts.[62] Also, Monsignor Nunzio Scarano, a senior official at the Vatican Bank who owns luxury properties and expensive works of art, is accused of transferring $26 million out of the Vatican Bank and smuggling it to Switzerland to help rich friends avoid taxes.[56] Cardinal Tarcisio Bertone, the Vatican's former secretary of state, is being investigated for steering $20 million (that was eventually lost) from the Vatican Bank to a film company run by a friend.[57] The extent of other corruption since the Vatican Bank's founding in 1942 is not known.

Jesus threw the moneychangers out of the temple. What would he do today?

Chapter 16

Child Sexual Abuse Cases

For the first thousand years of Church history, priests were allowed to marry. In 1139, celibacy for priests was instituted. So, for the following nine hundred years, some priests were celibate, some had girlfriends, some had boyfriends, and some sexually abused children.

Sexual abuse of children by the clergy (and in some cases by nuns) has been reported in countries throughout the world.[63, 64] (And there have also been cases of sexual abuse of nuns by priests.)[65]

These sexual predator cases involved the clergy who directly committed these acts, and the bishops who covered up the crimes by moving pedophile priests from parish to unsuspecting parish (and sometimes to other countries, according to a 2004 report by the *Dallas Morning News*). Many of the sexual abuse cases span several decades.

In the United States, the John Jay Report of 2004 said that:

- 81% of the victims were male,
- 19% were female,
- 22% were ten years of age or younger,
- 51% were eleven to fourteen years of age,

- 27% were fifteen to seventeen years of age,
- 25% of allegations were made within ten years of the incident,
- 50% were ten to thirty years after the incident, and
- 25% were more than thirty years after the incident.

In the United States, the Church from 1950 to 2012 paid an estimated $3 billion in settlements involving more than 3,000 priests. In 2003, John L. Allen, Jr., senior correspondent for the *National Catholic Reporter,* said while the US Conference of Catholic Bishops called for swift and sure punishment of guilty priests, the Vatican was more concerned with everybody's rights being respected, including the accused clergy, saying that media coverage was "unfair, excessive, and disproportionate."

These crimes have been brought to the forefront and talked about relatively recently. We have no reason to not believe that these sexual abuse crimes have gone on for two thousand years, not to mention child physical and mental abuse crimes.[66]

Chapter 17

The Church's Treatment of Women

The Church is always centuries behind the public's general acceptance of new ideas. Its doctrines and attitudes toward human sexuality and contraception are no exception and are archaic and nonsensical. Its opposition to contraception has doomed millions to unwanted pregnancies, overpopulation, poverty, and disease. It even opposes condom use in the fight against HIV/AIDS. Even if a married person's spouse is infected with HIV, the Church bans the use of condoms. How many have died because of this policy? It also opposes in vitro fertilization for couples who can't conceive through ordinary means.

Its views toward women go back two thousand years, and, while Catholic women don't wear burkas, they're treated as second-class citizens. It's a form of abuse that you would normally expect in the Third World. (Eventually, with economic development and increasing freedom, Muslim women will be free of the burka, just as many Catholic nuns in the United States no longer wear the habit.)

The Church says that women can't be priests because Jesus was a man and chose men to be his apostles; therefore, the Church is bound by Jesus's choice. But in Jesus' day, the roles of men and women were very different. Today, women routinely do jobs that long ago were considered "male" jobs—they're doctors, politicians, business leaders, and soldiers. The same can be said of men—some are childcare providers, elementary school teachers, and nurses.

If Jesus came here today and picked new apostles, most of us would believe that some would be women. (Even the ancient Egyptians had female priests.) In fact, for the Church to be technically correct, only male Jews located in the general area of today's Israel/Palestine should be priests since that's who Jesus and the apostles were. So, the Church's baffling, insincere, convoluted, and corrupt logic to exclude women is nonsense. (And by the way, Jesus didn't speak Latin.)

In some ways, it's understandable that a group of old men want to keep their exclusive club "for men only" as has happened in business, golf clubs, fraternal groups, etc. But even at Augusta National and St. Andrews Golf Clubs, the barriers have slowly come down. Not so the Church. The real tragedy is that women lack the courage to unite and boycott their Church. If they did, this policy would be changed overnight. Why don't they do that?

The bottom line is those who choose to believe that Jesus is God don't need the Roman Catholic Church in order to follow his teachings.

Chapter 18

Crimes of Other Religions

Due to its limited scope and the author's background, this book has focused on Catholicism and to some extent Judaism. (And, in general, if you discredit one religion, you've discredited them all.) Other religions are just as guilty of despicable crimes, like the women in India and Pakistan killed in "honor" killings.

- Jews kill Muslims.
- Muslims kill Jews.
- Hindus kill Muslims.
- Muslims kill Hindus.
- Sunni Muslims kill Shia Muslims.
- Shia Muslims kill Sunni Muslims.
- Sikhs kill Hindus.
- Hindus kill Sikhs.
- Buddhists kill Hindus.
- Hindus kill Buddhists.
- Christians kill Muslims.
- Muslims kill Christians.
- Christians killed Taoists.
- Taoists killed Christians.

- Catholics killed Protestants.
- Protestants killed Catholics.

Religion has gradually lost its political power in the West, greatly diminishing religious horrors. But religion remains politically strong in the Third World, where religious killings still run rampant. (Eventually, the poverty, lack of education, and lack of opportunity will ease, and progress will eliminate the radicals of Islam.)

So, what intelligent person with a good heart would ever belong to one of these religions?

Chapter 19

The Church's Suppression of Knowledge

The decline of the Roman Empire starting around 376 AD resulted in much medical and scientific knowledge from ancient Greece and Rome being lost or destroyed.

At the same time, the power and influence of the Roman Catholic Church was rising, and politics and beliefs were soon dominated by the Church. The Church was hostile to general learning by the public. Only the clergy were allowed to read the Bible, and science was of little use since "the Bible had already revealed all the truths that a person needed." The Church said that illnesses were sent from God to those who sinned (or were from demons and devils) and told the people to pray to the saints for cures, who would in turn appeal to God. (This is the same approach the Egyptian priests used thousands of years earlier.) The Church restricted science and medicine. Corpses could no longer be dissected, stunting medical research. The Church's authoritarian system stifled the creativity needed for scientific advancement and disagreement with Church rules was heresy.

Here are some examples of the Church's suppression of new knowledge. Remember, the Church at the time was one of the main keepers of historical records, so we don't know how many other people throughout history were persecuted.

Arnaldus de Vila Nova was an alchemist, astrologer, physician, and teacher who contributed to chemistry, medicine, and physics. He translated medical texts from Arabic and authored many medical writings. He was a pioneer in anesthesia. His skill in chemistry led to rumors of his involvement with the devil, and in 1299, the Inquisition ordered the burning of his writings.[67]

Cecco d'Ascoli was a mathematician, astrologer, and poet. He was a free thinker who theorized that people lived on the other side of the world. For attempting to use the stars to determine Jesus's birthdate, he was burned at the stake in 1327 at seventy years old.[15]

The Black Death, so called because a person's skin would blacken and because of a fearsome mortality rate, struck Europe in 1347. The disease was carried by fleas living on rats, and in a few years, it killed one-third of the population of Europe—twenty-five million people.[68] Sporadic outbreaks continued to kill people for decades, and it took 150 years for Europe's population to recover from one of the most deadly disease outbreaks in history. Adhering to Church beliefs, some people thought that their sins had caused the outbreak, and others whipped themselves as punishment ("flagellants").

Medical knowledge at the time was not capable of dealing with the plague. The question is how many lives could have been saved if the Church hadn't stunted the advancement of medicine for a thousand years from 500 AD to 1500 AD; how much blame for these deaths can be attributed to the Church? How many millions may have been saved? (Some think that the Church's persecution of witches led to widespread killing of cats, thus allowing rats to run free, spreading the plague.)[6] Historian Richard Carrier says that the Church halted the development of science, and, if not for that, our knowledge and technology today would be a thousand years more advanced.[69]

Ramon Llull was a prolific writer, philosopher, and artist and was a pioneer in mathematics and statistics as well as architecture. He had many of his books and theories condemned by the Inquisition in 1376.[70]

Andreas Vesalius was a skilled anatomist and physician who dissected cadavers and published his findings. He is considered the founder of the study of human anatomy. In 1543, he was persecuted by the Church and sentenced to do a pilgrimage to the Holy Land, during which he died at age fifty.[71]

Michael Servetus was a pioneer in the study of the circulation of blood. Unfortunately, he also published books critical of Church doctrine, including the Trinity. In 1553, he was imprisoned by the Church and his books burned. He escaped to Switzerland and was captured and burned at the stake by Calvinist Protestants.[72]

Tommaso Campanella, a philosopher, was tortured several times by the Inquisition starting in 1594 for his unorthodox views and his defense of Galileo; he spent twenty-seven years in prison.[73]

Giordano Bruno, a philosopher, mathematician, and astronomer, thought that the sun was just a typical star, that planets orbited other stars, and that there were an infinite number of inhabited worlds. He also opposed several Church doctrines, and after a seven-year trial, was burned at the stake in 1600.[74]

Lucilio Vanini was a priest, philosopher, and writer who thought that animal and human souls were equal. In 1619, his tongue was cut out, and he was strangled and burned.[75]

Galileo was an accomplished mathematician, astronomer, physicist, and philosopher who made many contributions to science. In 1543, Nicolaus Copernicus said that the earth revolved around the sun, contrary to the belief of the day that the earth was at the center of the universe, and the sun revolved around the earth. Copernicus wrote in Latin, which relatively few people understood, and shortly after publishing his book, he died before the Church could get hold of him. In 1632, Galileo published a book in Italian (which many could read) defending Copernicus's idea that the earth revolved around the sun.

The Church, concerned about the spread of this idea, called him before the Inquisition. The Inquisition said that the idea of the earth revolving around the sun was foolish and contradicted the Bible. In 1633, at age seventy, he was found guilty of heresy, threatened with torture, forced to recant his theories, and sentenced to life in prison, which was commuted to house arrest at his farm, where he died in 1642.[76] His books were banned and stayed on the Church's Index of Prohibited Books until 1835.

Pope John Paul II ordered a reexamination of the evidence against Galileo and offered a form of an apology (saying that both the Church and Galileo were to blame) in 1992, some 360 years later.

Why did the Church suppress and persecute Galileo? Maybe because when the Abrahamic religions (Judaism, Christianity, Islam) were founded, the common belief was that the earth was a much smaller size than we know it to be today and was believed to be at the center of the universe. The sun and planets revolved around it, and the stars were just twinkling lights. This made the earth a unique, very important place. It also made the inhabitants of the earth (humans) very special and important, and it made the institutions (the Church) that ruled these inhabitants even more important. So, perhaps the Church leaders, being clever and savvy for their day, could see that if the earth was just another planet and humans were not so special, then they would ultimately lose their power and influence. And so, they cracked down on Copernicus's and Galileo's ideas, hoping to preserve their power.

Over time, knowledge won out over Church doctrine, and people came to understand that the earth revolved around the sun. And, perhaps, the surprising thing to Church leaders was that people still kept coming to Church, giving the Church money and control, even though the people realized that the earth is but a tiny speck in the vast universe.

Isaac La Peyere, a French theologian, published *Men before Adam* in 1655. In the book, he claimed that people existed before Adam. His books were burned, and he was briefly imprisoned.[75]

Georges-Louis Leclerc was a mathematician, scientist, and author who studied fossils and suggested that the earth was far older than stated in the Bible, and that life forms evolved over time. He was a predecessor to Darwin. He was persecuted and humiliated by the Church and forced to resign from Sorbonne University around 1778.[77]

Edward Jenner, a doctor, developed smallpox vaccination in 1796. His work may have saved more lives in the last two hundred years than anyone else. He was criticized by the clergy, who said it was ungodly to inoculate a person with materials from an animal.[68]

James Young Simpson, a physician, the first to use chloroform to ease the pain of childbirth, was attacked by the clergy, who said that a woman's pain during childbirth was punishment for the sins of Eve. They quoted Genesis 3:16 in which God says, "I will greatly multiply your pain in childbirth; in pain you will bring forth children; your desire will be for your husband, and he will rule over you."[78]

Charles Darwin published his theory of evolution in 1859. Evolution described how animals and plants change and evolve over time, eventually leading to new animals and plants, and that humans descended and evolved from apelike creatures. The Church vehemently opposed this line of thinking and declared that the Bible stated that God created man.[79]

Once again, as with Galileo, the Church leaders began to sweat bullets. If evolution created humans, then the Bible must be wrong, and if it's wrong, the Church could lose its power over the people.

The Church, 150 years after Darwin's theory, now admits that evolution is true—it's how humans came to be. To save face, Pope Pius XII in 1950 and Pope John Paul II in 1996 said that while evolution can explain the human body, the soul and the mind are a result of God.

As with Galileo, the Church had to back down again. (Today, in developed countries, knowledge is spreading, and Church power and control is waning. Some evangelical groups insist that the Bible be interpreted literally and that God created humans directly. The spread of knowledge will eventually catch up to them, and they'll either accept evolution as fact or be increasingly marginalized and thought of as crackpots.)

Georges Lemaitre was a Belgian priest and physicist who attended Cambridge, Harvard, and MIT and was familiar with the thinking of Albert Einstein (space and time) and Edwin Hubble (an expanding universe). In 1931, Lemaitre theorized that the universe began from an infinitely dense single point, from which space and time expanded rapidly, eventually becoming the universe that we know today. This big bang occurred over 13.5 billion years ago.

The Church, surprisingly, did not excommunicate or persecute Lemaitre, but their problems were growing bigger. Their belief that the earth was the center of the universe, which made humans and the Church extremely important, was shattered by Galileo in 1633. Their belief that God created humans directly was shattered by Darwin in 1859.

So Pope Pius XII in 1951 accepted the big bang concept of the universe beginning from a single point. But in a surprising display of illogical thinking, he said that the big bang proved that God exists—it did, of course, do nothing of the kind.[80]

Stephen Hawking, Leonard Mlodinow, Lawrence Krauss, Victor Stenger, Robert Matthews, and others, all expert physicists, say that research and development over the decades since 1931, including the constantly developing science of quantum mechanics (the physics of the very small), led them to theorize that the universe could have come into existence all by itself. They say there is no need for a God to explain the existence of the universe.

Now what does the Church do? What illogical psychobabble will it release in Latin to explain this? (Some may say that "they didn't know any better at the time." What about our time? Who do we persecute, ridicule, and discriminate against? In a thousand years, how will our beliefs and actions be judged?)

The Church has not been "enlightened" throughout its history despite its self-stated "closeness to God." How can this be? And it's not alone. Christian Scientists believe that illness is an illusion and thus deny proper medical treatment; untold thousands have died from this belief. Jehovah's Witnesses believe that blood is sacred and refuse blood transfusions. These are a few examples of religious nonsense.

Our Current God/Religion

The Church, over the years, may have been responsible—directly or indirectly—for more atrocities and deaths than Hitler. If Hitler had apologized for the Holocaust, would that have absolved him? If the Church apologizes for past sins, does that absolve the Church? If this is our current God/religion, do we really need a God?

"There is another form of temptation . . . this is the disease of curiosity . . . which drives us to try and discover the secrets of nature, those secrets which are beyond our understanding, which can avail us nothing and which man should not wish to learn."

—Augustine (354 AD–430 AD), saint

"Reason should be destroyed in all Christians."

—Martin Luther (1483–1546), priest

"You will notice that . . . the Church has always favored the doctrine which most completely subjugated the human mind and annihilated reason."

—Voltaire (1694–1778), French philosopher

"To command their professors of astronomy to refute their own observations is to command them not to see what they do see and not to understand what they do understand."

—Galileo Galilei (1564–1642), astronomer

"The Church says the earth is flat, but I know that it is round, for I have seen the shadow of the moon, and I have more faith in a shadow than in the Church."

—Magellan (1480–1521), explorer

"But how has it happened that millions of fables, tales, and legends have been blended with both Jewish and Christian revelation that have made them the most bloody religions that ever existed?"

—John Adams (1735–1826), second president

"Millions of innocent men, women, and children, since the introduction of Christianity, have been burned, tortured, fined, imprisoned."

—Thomas Jefferson (1743–1826), third president

"The Catholic Church is more concerned about preventing contraception than preventing child rape; it's more concerned about preventing gay marriage than genocide."

—Sam Harris (1967–), neuroscientist

"Man will never be free until the last king is strangled with the entrails of the last priest."

—Denis Diderot (1713–1784), French philosopher

"If you have a few hundred followers and you let some of them molest children, they call you a cult leader. If you have a billion, they call you pope."

—Bill Maher (1956–), TV host

"Men never do evil so completely and cheerfully as when they do it from religious conviction."

—Blaise Pascal (1623–1662), mathematician

"I cannot give liberty of conscience to the pope and his followers . . . so long as they tell me . . . that their conscience orders them to burn my wife, and strangle my children, and cut my throat when they find their opportunity."

—Abraham Lincoln (1809–1865), sixteenth president

Part V
Do We Need a God?

Chapter 20

What If People Didn't Believe in God/ Religion?

Would Good People Turn Bad?

If people didn't believe in God/religion, would murder and mayhem result? Would society fall apart?

We all know many church-going, good people, who, if they suddenly didn't believe in God/religion, would not become degenerates and murderers. They'd continue as good people.

At one time, it was thought that because humans used tools and language and formed social groups, it set them apart from animals. We now know that many animals use tools (in fact, female chimps are believed to have invented the first spears used in hunting[1]) and language and have social organization.[2]

Many insects and animals, such as ants, termites, bees, wasps, birds, dolphins, monkeys, cats, dogs, mice, elephants, horses, whales, hyenas, rabbits, lions, penguins, rats, and wolves exhibit social order and behavior. A vampire bat may regurgitate a small amount of blood to sustain another bat between nightly feedings.[3] Elephants honor their dead, nurse their wounded, communicate with their feet, and cry.[4] Ants organize themselves to find food, battle,[5] and build rafts and bridges of linked-together ants.[6] Dogs display signs of jealousy,[7] and, as every dog owner knows, they make an effort to listen to and understand words directed at them.[8] Rats display regret.[9] A female octopus guarded her eggs for fifty-three months at the expense of her own nutrition.[10] Primatologist Frans de Waal says that chimpanzees and other primates show empathy, have a sense of fairness, take care of those in need, share with those less fortunate, and feel guilt and shame. He says these actions are the building blocks of morality and are far older than humanity; morality comes from within, not above.[11] Neuroscientist Michael Gazzaniga says that our brains have built-in mechanisms for fairness, equality, trust, and altruism.[12] (And even sharks exhibit social personalities[13]).

Our brains evolved for social interaction with others, and we have a built-in social moral code to govern our behavior without the need for God/religion.

Ants

There are probably thousands upon thousands of ants in my yard. While I don't have the power to create them, I do have the power to disrupt their lives or destroy them. Do they think of me as a god? I crush them when I walk on the grass, decapitate and hurl hurricane-force winds at them with my lawnmower, and poison their friends when they try to sneak into my house looking for sweets. To them, I'm an Old Testament God. Do they have little altars and pray to me? Are they not intelligent enough to do that, or do they just go about their lives, living in their highly socialized and ordered community, not concerned with paying someone off to get them into Ant Heaven? And since an ant is more significant to the earth than a human is to the universe, isn't all this talk of God sort of nonsensical?

The Religion-Government Complex

In his farewell address, President Eisenhower in 1961 warned the country of the growing power and influence of the military industrial complex. Politicians approve military spending programs, and the military sends contracts and money into the politician's home community, generating jobs for the workers, who will, in turn, reelect the politicians. (This spending has mushroomed, resulting in the United States spending almost four times as much on the military as China, eight times as much as Russia, and eleven times as much as Great Britain, thus diverting funds from other needed areas.)

There's also a religion-government complex. Through most of human history, religion and government were intertwined.[14] Governments used religion to help manipulate and control the people. Surprisingly, even today, prayers in the US Congress, presidents invoking God's name in times of war or crisis, and slogans like "One Nation Under God" continue in our "advanced" society.

English philosopher John Locke (1632–1704) argued that government should have no authority over individual conscience. Thomas Jefferson was a strong proponent of the separation of church and state. The US Constitution and the Bill of Rights guarantee freedom of religion without government interference, and the separation of church and state continued with the French Revolution.

Today, in advanced countries, church and state are separate. But in backward regions like the Middle East, government and religion are still intertwined.

Years ago, the nuns would tell us to "offer up our sacrifices and disappointments" and, in exchange, we would receive our reward in Heaven. But what if people didn't believe in God/religion—especially the poor and disadvantaged? If there was no reward awaiting in Heaven, then the poor would expect more out of life on this earth and pressure governments to provide it. Governments cozy up to religion, using it as a tool to keep the disadvantaged under control (by promising them a better life in Heaven), thereby keeping the pressure off and preventing revolutions.

"Religion is regarded by the common people as true, by the wise as false, and by rulers as useful."

—Seneca (4 BC–65AD) Roman philosopher

"Religion is excellent stuff for keeping common people quiet. Religion is what keeps the poor from murdering the rich."

—Napoleon (1769–1821), French leader

Imagine No Religion

If we didn't have God/religion, would people kill each other because of country, tribe, skin color, gender, or eye color? We don't know. But we do know that they do kill each other in the name of their God/religion.

"Imagine there's no Heaven."

—John Lennon (1940–1980), musician, *Imagine*

"Is God willing to prevent evil, but not able? Then he is not omnipotent. Is he able, but not willing? Then he is malevolent. Is he both able and willing? Then whence cometh evil? Is he neither able nor willing? Then why call him God?"

—Epicurus (341 BC–270 BC), ancient Greek philosopher

"I have recently been examining all the known superstitions of the world, and do not find in our particular superstition (Christianity) one redeeming feature. They are all alike, founded on fables and mythology."

—Thomas Jefferson (1743–1826), third president

"The age of ignorance commenced with the Christian system."

—Thomas Paine (1737–1809), founding father

"The Bible is not my Book and Christianity is not my religion. I could never give assent to the long complicated statements of Christian dogma."

—Abraham Lincoln (1809–1865), sixteenth president

"This would be the best of all possible worlds, if there were no religion in it."

—John Adams (1735–1826), second president

"Religion I found to be without any tendency to inspire, promote, or confirm morality, serves principally to divide us and make us unfriendly to one another."

—Benjamin Franklin (1706–1790), inventor and statesman

"For me the Jewish religion like all other religions is an incarnation of the most childish superstitions."

—Albert Einstein (1879–1955), physicist

"Religion is an insult to human dignity. With or without it you would have good people doing good things and evil people doing evil things. But for good people to do evil things, that takes religion."

—Steven Weinberg (1933–), physicist

"So far as religion of the day is concerned, it is a damned fake. . . Religion is all bunk."

—Thomas Edison (1847–1931), inventor

"I believe in God, only I spell it Nature."

—Frank Lloyd Wright (1867–1959), architect

"Religion began when the first scoundrel met the first fool."

—Voltaire (1694–1778), French philosopher

"All religions bear traces of the fact that they arose during the intellectual immaturity of the human race—before it had learned the obligations to speak the truth."

—Friedrich Nietzsche (1844–1900), German philosopher

"In spite of all the yearnings of men, no one can produce a single fact or reason to support the belief in God and in personal immortality."

—Clarence Darrow (1857–1938), lawyer

"Religion is an illusion and it derives its strength from the fact that it falls within our instinctual desires."

—Sigmund Freud (1856–1939), neurologist

"The Christian God is a being of terrific character—cruel, vindictive, capricious, and unjust."

—Thomas Jefferson (1743–1826), third president

"If there are any gods whose chief concern is man, they can't be very important gods."

—Sir Arthur C. Clarke (1917–2008), science fiction author

"All religion, my friend, is simply evolved out of fraud, fear, greed, imagination, and poetry."

—Edgar Allan Poe (1809–1849), author

"Civilization will not attain to its perfection until the last stone from the last church falls on the last priest."

—Émile Zola (1840–1902), French author

"Stripping away the irrational, the illogical, and the impossible, I am left with atheism."

—Mark Twain (1835–1910), author

"Religion is . . . the opium of the people."

—Karl Marx (1818–1883), German philosopher

"It is difficult to imagine a set of beliefs more suggestive of mental illness than those that lie at the heart of many of our religious traditions."

—Sam Harris (1967–), scientist

"I believe that religion, generally speaking, has been a curse to mankind—that it's modest and greatly overestimated services on the ethical side have been more than overcome by the damage it has done to clear and honest thinking."

—H. L. Mencken (1880–1956), journalist

"Christianity indeed has equaled Judaism in the atrocities, and exceeded it in the extent of its desolation. Eleven million men, women, and children have been killed in battle, butchered in their sleep, burned to death at public festivals of sacrifice, poisoned, tortured, assassinated, and pillaged in the spirit of the Religion of Peace, and for the glory of the most merciful God."

—Percy Bysshe Shelley (1792–1822), poet

"We are all atheists about most of the gods that humanity has ever believed in. Some of us just go one god further."

—Richard Dawkins (1941–), biologist

"There was a time when religion ruled the world. It is known as the Dark Ages."

—Ruth Hurmence Green (1915–1981), author

"The fact that a believer is happier than a skeptic is no more to the point than the fact that a drunken man is happier than a sober one."

—George Bernard Shaw (1856–1950), Irish playwright

"Life in Lubbock, Texas, taught me two things. One is that God loves you, and you're going to burn in Hell. The other is that sex is the most awful, dirty thing on the face of the earth, and you should save it for someone you love."

—Butch Hancock (1945–), songwriter

"Suppose we've chosen the wrong god? Every time we go to church, we're just making him madder and madder."

—Homer Simpson (1987–), TV character

"The only good thing to come out of religion was the music."

—George Carlin (1937–2008), comedian

"Religion. It's given people hope in a world torn apart by religion."

—Jon Stewart (1962–), political satirist

God Not Required

So, if we don't need a god to explain our, or the universe's, existence, and if society would continue to function without God/religion, why do we believe?

Part VI
Believing

Chapter 21

Why Do We Believe?

Over 90 percent of Americans believe in God, according to the AP/GFK poll of March 2014 and the Gallup poll of May 2011.[1] And 70 percent believe in angels, 60 percent in the devil, 42 percent in ghosts, 36 percent in UFOs, and 26 percent in witches, according to Harris polls in 2005 and 2013. Whether these statistics are totally accurate or not, it is probably true that most people believe in God.

Why do people say they believe? They say they believe in God because of the complexity found in a maple leaf or a butterfly or the human brain or the universe. Or because it gives meaning and purpose to life. Or because they find it comforting. Or they need to believe in something, because of the fear of death, the fear of nothingness. Or because the Bible says so, or they've had a "God experience."[2]

But why, in fact, do they believe?

Programming

Today, before I make any big decision, I question, research, and gather information. Then, after pondering, I decide what to believe. So, I question first and then believe.

But with religion/God, when we were young, we "believed" first, and now, as we grow older, we question.

Very few of us chose our God/religion. Most of us were indoctrinated, instructed, and programmed by our parents, churches, and society at a very early age.[3] And programming, especially at an early age, can be very persuasive and controlling.

Is all programming bad? No. Whenever I rarely have a glass of milk, I finish it because my mother told me to "finish your milk." I was also programmed to say "please" and "thank you," hold open doors for women, not take rides from strangers, work hard, be on time, help the less fortunate, and hundreds of other things that help me navigate through life. (I was also programmed to not write this book—my mom would be worried for me; my dad would be a little puzzled but silently smile.)

Taken to extremes, programming can have a dark side, like covering up pedophile priests or the control that some cults have over individuals.

The Watch Tower Woman

A quiet tapping on my car window mildly startled me one cold, bleak winter day as I sat at lunchtime reading a book at the beach.

It was a nondescript middle-aged woman whom I perceived as alone and lonely. She gently offered me a copy of *The Watchtower*, the magazine of Jehovah's Witnesses. I smiled and politely declined. (Several times, over the years, when Jehovah's Witnesses came to my house, I politely declined their overtures but deep down felt sorry for them, wanted to engage them, shake them, and help them snap out of it.)

She smiled and asked if I wanted to know more about God. I softly told her that I didn't believe in her kind of God. She continued with the meek, robotic smile, aware that she had come upon a soul that needed saving. She said that the Bible was "the word of God."

Somewhere in my mind, the idea for this book was beginning to form; I responded that the Bible was written by men, not women, thousands of years ago and was "the word of men" not God.

She persisted, saying, "No, it was the word of God."

Now, I kindly, and a little firmly, told her that she had been brainwashed and that she should use the brain that her God had given her to figure things out for herself. She grew frustrated, realizing that the soul she had hoped to save was beyond reach—she had failed.

She parted, saying, "There was no changing your mind, was there?"

I said no and again encouraged her to think for herself.

This poor waif of a woman, dourly clad and seemingly alone in the world, unable to think for herself and taken advantage of by a cult of mindless robots, slowly walked away.

She will never read this or other books with alternative views, and she may never find the courage to face life by herself and seek freedom from the cult that controls her.

Astoundingly sad—the zombies do truly walk among us.

So after our initial programming as children, why do we continue to believe as adults?

Tradition

As adults, most of us continue with the God/religion in which we were raised because of tradition.[4] Our grandparents, parents, siblings, relatives all have the same beliefs (although Uncle Charlie may be straying). We gather on religious holidays, have enjoyable, meaningful times, and carry on the belief, the tradition.

"Tradition means giving votes to . . . our ancestors. It is the democracy of the dead."

—G. K. Chesterton (1874–1936), author

"A love of tradition has never weakened a nation . . . but the new view must come, the world must roll forward."

—Winston Churchill (1874–1965), politician

Hope

We hope for good weather, we hope for a raise, and we hope our kids are okay.

Each spring, I excitedly head to the local golf superstore, try out every putter, and always buy one. Why? Hope. I know and hope that this new putter will dramatically improve my game.

Hope is very powerful; hope is very seductive.[5] Hope invades our minds daily, continuously, relentlessly.

Not a day goes by in the news that someone isn't swindled, defrauded, and taken advantage of because they "hoped." There's the Nigerian scam promising a big share of entangled funds that can be freed up if you provide a small amount of money and your banking information. There's the TV scam at the mall. The caller tells the victim that there's a surplus of big screen TVs that can be bought for half price. The victim proceeds to the mall, gives the money to the thief in the parking lot, and receives a realistic-looking receipt. He's instructed to go around to the back of the store to the shipping door and wait. The thief heads to the store's front doors, and when the victim is out of sight, pockets the money and disappears. After a good period at the shipping door, the victim calls the cops.

There's the grandparent scam, where the caller pretends to be a friend of a grandson who's in jail and needs the grandparents to wire transfer money to the "friend" to bail him out. There's government-approved "hope machines" like the lottery and gambling casinos or selling people on home mortgages that they can't afford. Many of these schemes take advantage of the poor, elderly, and unsophisticated.

But the wealthy and "sophisticated" were massively victimized to the tune of billions of dollars by Bernie Madoff, who promised them optimistic financial returns.

So, wouldn't it be terrific if there were a Heaven? Let's just keep on hoping.

> "Hope in reality is the worst of all ends because it prolongs man's torments."
>
> —Friedrich Nietzsche (1844–1900), German philosopher

> "Hope springs eternal."
>
> —Alexander Pope (1688–1744), poet

Inertia

Laziness.

Lack of curiosity.

"Let's just go along."
"It's easier."
"I'm too busy to bother."
"What does it matter?"
"Let God do it."

"The inertia of the mind urges it to slide down the easy slope of the imagination, rather than to climb the steep slope of introspection."

—Marcel Proust (1871–1922), author

"Nothing happens until something moves."

—Albert Einstein (1879–1955), physicist

Fear

Fear of death, fear of Hell, fear of nothingness—all these drive belief.[3] Buddhists say that belief in God is just a response to fear.[6]

"We can easily forgive a child who is afraid of the dark; the real tragedy of life is when men are afraid of the light."

—Plato (428 BC–348 BC), Greek philosopher

> "Now is the time to understand more, so that we may fear less."
>
> —Marie Curie (1867–1934), physicist

Comfort

Belief in God/religion may have made our very ancient ancestors better organized and productive while making them less fearful of death and other things they didn't understand.[5] Today, God/religion provides comfort to some in times of stress and crisis—they don't feel alone.

Many think that it doesn't hurt to believe, and it would be nice if it (God) were true.

> "If you look for truth, you may find comfort in the end; if you look for comfort, you will get neither truth nor comfort."
>
> —C. S. Lewis (1898–1963), author

Mental Illness

Unfortunately, some people suffer depression, experience hallucinations, and delusions, or other forms of mental illness. They talk to God, God talks to them, they believe in magic, and don't deal effectively with reality.[7]

To believe in God/religion doesn't mean that you have mental illness. It's just that some unfortunates take it to extremes.

Not Using the Brain/Lack of Courage/Irresponsibility/Lack of Self-Belief/Ostrichitis

Like the ostrich with its head in the sand, lack of courage, lack of confidence, and irresponsibility will cause people to not use their brains.[3, 4, 5]

Rather than face the possibility that God may not exist, some will run and hide—they'd rather be blissfully ignorant than shake their foundations.

"It takes courage to grow up and become who you really are."

—E. E. Cummings (1894–1962), poet

"No one can make you feel inferior without your consent."
—Eleanor Roosevelt (1884–1962), politician

"Most people are frightened of responsibility."
—Sigmund Freud (1856–1939), neurologist

Conformity

Some people believe that the doctor's always right, as is the teacher, police officer, banker, priest, lawyer, politician, journalist, and TV commentator. When we were young, our programming required us to believe authority figures, do as we were told, and follow the leader.

While we need leadership and governance, many people too readily conform to the wishes of the group or feel that in carrying out another person's instructions, they no longer are responsible for their own actions.[8] (Is this what happened to ordinary Germans in their involvement in the Holocaust?)

Anarchy has no place, but it's every individual's inalienable right and duty to question authority figures.

Why do people refuse to think for themselves? Do they lack curiosity? Why are there so many robots and sheep? Do most people just like to follow orders? Why do people hide from reality? Does everyone hide a little/a lot, depending on the subject matter? Is denial of reality the cousin of hope?

Why are people afraid to question the existence of God?

Questioning

My friends and I often discuss and exchange books, and at a recent meeting, I brought along my just-read copy of Christopher Hitchen's *God Is Not Great—How Religion Poisons Everything.* The title sums up the book. When I arrived, I asked the group of six if any would like to read it.

- Two said yes
- Two said nothing
- Two blankly stared

A few hours later, when I was leaving, I asked the two who said yes initially if they wanted the book. They politely declined.

Were all six:

- So sure in their God beliefs that they thought that the book was sacrilegious or silly?
- Reluctant to do the work, the thinking?
- Truly not interested in the subject matter?

- Afraid to question, and, if so, where does this fear come from—fear of an idea, a concept?
- Just oblivious, content to just float along through life?

Why are we here? What is our purpose? What's our ultimate destiny? These questions have inspired and fueled philosophers for thousands of years. Why aren't people interested in pursuing these questions? While it's not our responsibility to dictate our own answers to these questions, it is our responsibility to gently raise them and get the questioning process going.

Could "Belief" in Something That Doesn't Exist Be Harmful?

Freud called religious belief an illusion and thought that it contributed to a weakness of the intellect by shutting off curiosity and thinking.[9] Richard Dawkins calls belief in God a delusion.[10]

But if belief in God/religion helps comfort people, is it harmful?

"Belief" can be harmful if:

- it takes time, money, and resources from you, that could be better invested in your loved ones and the needy;
- it causes you to neglect your responsibilities and your family;
- it hypnotizes you and prevents you from demanding better living conditions, education, health care, housing, and job opportunities in this life and commands you to "accept your lot in life";
- it convinces you to neglect proper medical care;
- it dictates your daily actions and life;
- it requires you to discriminate against, persecute, or kill believers of other religions (as the Church did hundreds of years ago);

- it makes you elevate charlatans and con artists to positions of respect;
- it gives control of you to a cult;
- it forces you into a subservient position, allowing others to take advantage of you and your resources;
- it generates false hopes;
- it locks up your mind and curiosity;
- it denies you the opportunity of defining your own purpose; and
- it commands you to deny reality?

Critics[11,12] of twelve-step programs (Alcoholics Anonymous, Narcotics Anonymous, Gamblers Anonymous, etc.) say that the programs have very low success rates and that other approaches, including no treatment, acupuncture, and exercise, are more effective. One reason for this may be that AA members must submit to the will of God, taking away their personal control and responsibility, while making God responsible for recovery. If instead the members were forced to accept that they were fully responsible, would that improve success rates?

Is excess alcohol or drug use harmful if it makes people happy, content, satisfied? An addict avoids the pains of life through heroin, cocaine, alcohol, and OxyContin. So, the "belief" in God/religion/Heaven is a "Novocaine" of the mind, which dulls the pain of reality, the pain of our ultimate demise into nothingness, and of the realization that there is no eternal life. "Belief" is a societal-approved "drug," it's a (white) lie, an escape, an avoidance, a denial—it's an approved "addiction."

Heroin, cocaine, OxyContin, and "belief" are all harmful. They prevent people from striving to fulfill their capabilities or seeking what's owed to them in this life. With "belief," people pass off their own weaknesses, inadequacies, and laziness to God rather than taking responsibility themselves. “Belief” produces robots and zombies, not to mention the money, time, and resources that could be redirected to the poor and less fortunate.

Of course, “belief” is harmful.

If you wouldn’t do drugs, then why do you “believe”? And if you “believe,” what proof do you have that God exists?

> “It is a remarkable coincidence that almost everyone has the same religion as their parents and it always just so happens it’s the right religion.”
>
> —Richard Dawkins (1944–), biologist

> “You can’t convince a believer of anything; their belief is not based on evidence but a deep-seated need to believe.”
>
> —Carl Sagan (1934–1996), astrophysicist

> “I call him free who is led solely by reason.”
>
> —Baruch Spinoza (1632–1677), Dutch philosopher

> “Faith means not wanting to know what is true.”
>
> —Friedrich Nietzsche (1844–1900), German philosopher

> “Faith is believing what you know ain’t so.”
>
> —Mark Twain (1835–1910), author

> “Don’t ya know what faith is? Faith is when you believe somethin’ that nobody in their right mind would believe!”
>
> —Archie Bunker (1971–1983), TV character

Part VII
Proof

Chapter 22

Talking With God

All bosses occasionally show up. Today, it's relatively easy for bosses to communicate by video and phone conferencing, TV, various printed publications, Facebook, Twitter, texting, and other social media. That's what bosses do. So where is this Ultimate Boss of Bosses? Where is this God? If He cares about us so much, and wants us to adore Him, couldn't He at least send us a tweet? He can't be too busy to communicate with us because there's no such thing as "too busy" for a God—He has infinite time and abilities. And even if I'm reluctant to use some of these communication methods, He's perfectly capable of using them all.

In the history of the world, over a hundred billion humans have been born. Of these, whom has He supposedly communicated with?

Long Ago

Thousands of years ago, God spoke/appeared to Moses, Abraham, Isaac, Jacob, and others, according to the Bible,[1] and pharaohs and ancient priests commonly talked to and heard from god(s). Talking to god(s) was not unusual at the time. The people of the time were very susceptible to superstition and magic, and there was so much that they didn't know that they believed that god(s) were responsible for every aspect of daily life.

Today

God supposedly spoke to Joseph Smith, who founded the Mormon Church; Ellen White, who founded the Seventh Day Adventist Church; Mary Baker Eddy, who founded the Christian Science Church; Sun Myung Moon, who founded the Unification Church; and Mike Bickle, who founded the International House of Prayer.[2] (Why did God instruct them to form five different religions? Shouldn't they all be the same? Are four of these religions wrong? Are all of them nonsense?) God supposedly talks to the faith healers and prosperity preachers, as well. And He is said to talk to many more.

So we have a collection of superstitious old men thousands of years ago (who thought the earth was flat) and deluded fanatics or swindlers/con men/women today who say that God communicated with them. (Today the clever preachers know that saying they "talked with God" enhances their influence, power, and wealth; the naïve preachers lack courage to face the truth.)

If I told my friends that God had appeared or spoke to me, they would, of course, laugh. If I told some strangers, most would immediately dismiss me; the rest would ask for evidence or proof. They'd want to listen to a recording of God's voice or see a picture on my cell phone. Of course, like all those previously mentioned, I have no evidence, I have no proof.

"Faith" is not proof.

So, let me see if I understand this . . . God only speaks to people in their minds—He doesn't telephone, text, tweet, e-mail; appear on radio, TV, or the Internet; and definitely doesn't appear in person to groups of individuals who would be able to corroborate His appearance.

So people hallucinate, or desperately, hopefully imagine, or dream up or make up God.

God has never shown up because He either doesn't care . . . or He doesn't exist. Why do we have difficulty understanding this?

Chapter 23

Santa

As a child, I fervently, excitedly believed in Santa.

If someone had asked me at that time how I knew that there was a Santa, I would have pointed to the toys. That was my proof—accompanied with the belief of my friends, the widespread endorsement of Santa in our culture, and the reinforcement of Santa by my parents.

I believed, I had faith, but that didn't make it true.

As I grew older, I lost the belief due to the rumblings of my friends and my acquisition of new knowledge/common sense. (How could a Santa deliver all those toys in a small sleigh one night all over the world and how could reindeer fly?)

In subsequent Christmases, the toys still came, but I knew that they came from my parents.

So, what exposed Santa was the toys—the physical evidence. If instead, Santa never brought toys but did magically visit every little girl and boy on Christmas Eve and gave them a blessing, making them better people, then, even today as an adult, that myth might still persist. He might be referred to as St. Nicholas just like all the other "saints."

The burden of proof falls upon those who believe in Santa to prove that Santa exists; it's not on those who never believed to prove that Santa doesn't exist. So with God. Where's the proof? Without the proof, we must assume that God does not exist.

When I was young, I believed in Santa. "Belief" doesn't cut it; "belief" is not proof.

Chapter 24

God's Attributes and Inadequacies

God's Attributes

God is said to be omnipotent (all-powerful), omniscient (all-knowing), and omnipresent (present everywhere at the same time). He is also said to be benevolent, loving, and holy (incapable of sin).[3, 4]

So, if God is all these things:

- Why is there evil?
- Why is there famine and starvation?
- Why is there disease?

If God is holy (incapable of sin), why did He kill so many people in the Bible? Is that loving? Is that merciful? It is said He is jealous and wrathful—not very becoming, in fact, downright sinful.

Darwin said he could not see a god in the pain and suffering inflicted by wasps that paralyze caterpillars, insert their eggs, and upon hatching, use the caterpillar as live food.

God's Insecurity

Q: Why Did God Make You?

A: God Made Me To Know Him, To Love Him, And To Serve Him In This World, And To Be Happy With Him Forever In The Next."

—*The Baltimore Catechism*

If I spend this life, knowing Him, loving Him, and serving Him, that's a lot of work. Why the hell would I want to do the same in the next?

While my kids are wonderful, I never had them with the thought that they would "know, love, and serve" me. How insecure is this God that He needs that kind of attention? That he needs a slave? How narcissistic. If He's all powerful, then He shouldn't need all this worship and adoration.

Miracles and Prayers

Bill Arpano

I met Bill Arpano in engineering school. He was a terrific guy—president of his high school class, warm, friendly, funny, smart.

He was also a Christian Scientist. Thirty years after graduating, he contracted prostate cancer. When I heard, I called him. He sounded upbeat; he was a year and a half into his battle with cancer. We talked. I told him that if it were me, I'd use every weapon I could get my hands on—radiation, chemotherapy, and anything and everything else. He was only using prayer. A short time later, he did begin radiation and chemotherapy, but it was too late. The cancer had metastasized—he died.

Here was a terrific guy, trained in engineering school to logically analyze problems. His parents programming overruled his logic and killed him.

Mary Sullivan

My cousin, Mary Sullivan, was a bubbly, wonderful person who dismissed heartache in her life, kept up a positive spirit, and worked for Catholic Charities (an organization that does good work). She developed breast cancer. She rejected medical treatment and prayed instead; she died. The programming killed her as well. (Can her children sue the Church for causing her to forego medical treatment?)

Q: Does God Know All Things?

A: God Knows All Things, Even Our Most Secret Thoughts, Words, And Actions.

—Baltimore Catechism

Q: Can God Do All Things?

A: God Can Do All Things And Nothing Is Hard Or Impossible To Him.

—Baltimore Catechism

Q: Is God, Just, Holy And Merciful?

A: God Is All Just, All Holy, All Merciful, And He's Infinitely Perfect.

—Baltimore Catechism

So, Bill and Mary prayed to God seeking a cure, seeking a miracle.[5] God, since He knows all things, heard their prayers. And since He can easily do all things and is merciful, He could have saved them. He didn't.

In the old days, the nuns would say "God works in mysterious ways," which, of course, doesn't make sense.

In 1988 Randolph Byrd, using 393 patients at the San Francisco Hospital coronary care unit, conducted a study to see if prayer could help any of these patients.[6] It didn't. The Mayo Clinic in 2001 studied the effect of prayer on 799 coronary surgery patients.[7] Prayer had no effect. In 2005, a three-year study by Duke University of 748 cardiology patients showed that prayer had no effect.[8]

In 2006, Harvard professor Herbert Benson conducted a prayer study of 1,802 coronary artery bypass surgery patients.[9] Prayer didn't work.

Apparently, the faithful have trouble distinguishing coincidence from reality. They remember and tout the one time prayer worked (by coincidence) and "forget" the thousand times it didn't.

The same is true of miracles. Those claimed in the Bible can and have been easily explained and refuted. But again, the people (Moses) in biblical times had very little knowledge. The rising sun was a "miracle" to them. They thought that all kinds of occurrences were miracles. Today, the Church "saints" someone if it can be shown that they were responsible for two miracles. The miracles are usually attributed to an ill person praying to a prospective saint for a cure. If the ill person gets better, "it must have been a miracle." Total nonsense.

Unfortunately, Bill and Mary weren't aware of the results of these prayer studies. If they were, maybe they would have taken appropriate medical action and still be alive today.

So, we have two possibilities.

Possibility A. We have a murderous God who slayed millions according to the Bible; who's incompetent and uncaring; and who likes to play cruel, sadistic games deciding who gets cured and who doesn't. If he was all loving, wouldn't he cure everyone? If you were God for a day, wouldn't you?

Possibility B.

Q: What Is Man?

A: Man Is A Creature Composed Of Body And Soul, And Made To The Image And Likeness Of God.

—*Baltimore Catechism*

If you and your friends got together and decided to invent a god, you would of course say that the god made you and your friends to his own image and likeness; it would make you feel good. So all the attributes previously described aren't God's attributes; they're human attributes.

Men made up the God attributes; men made up the God.

"I cannot imagine a God . . . made happy by my getting down on my knees and calling him great."

—Susan B. Anthony (1820–1906), social reformer

"Something is wrong. War, disease, death, destruction, hunger, filth, poverty, crime, torture, corruption, and the Ice Capades. If this is the best God can do, I am not impressed. This is not what you expect to find on the resume of a supreme being. It's what you expect from an office temp with a bad attitude."

—George Carlin (1973–2008), comedian

Chapter 25
Made-Up Stuff

Virgin Birth

When Mary gave birth to Jesus, she was a virgin. The Holy Spirit made her pregnant.[10]

The gospels tell the story of Jesus. They are positive stories about Jesus rather than historically accurate accounts of his life. Jesus died around 30 AD. The Gospel of Mark was written forty years later, and the gospels of Matthew and Luke, drawing heavily on Mark, were written ten and fifteen years after Mark. These three gospels tell a very similar story that historians regard as accurate. The gospel of John, written in 100 AD, is regarded by historians as a sales promotion story of Jesus with less historical accuracy.[11] Other gospels were written at the time, but the Church limited the officially approved gospels to those four to exert control over the Jesus story.

Very early accounts of Jesus do not mention a virgin birth. Matthew and Luke do mention it, but John and Mark do not. (Some say that the word *virgin* in those times didn't mean what it means today but rather "a young woman of childbearing age who had not yet given birth and who may or may not be a virgin.") So the reason that John and Mark don't mention the virgin birth may be that, in those days, thousands of years ago, miraculous births, including virgin births, were thought to be common occurrences among historical figures and gods.[12,13]

Kings and pharaohs in ancient times were thought to be divinely born. Pagans believed in virgin births.

In 3228 BC, Krishna in India was born to his mother, Devaki, who conceived miraculously. In 3000 BC, Isis, in Egypt, restores her dead husband, Osiris, to conceive her son Horus. Around 2500 BC, Melchizedek was born of a virgin, Sofonim. Around 2000 BC, Abraham's wife, Sarah, at age ninety, gives birth to Isaac. Around 1700 BC, the god Hercules was said to be born from the union of the god Zeus and the mortal woman Alcmene. The god Pan resulted from the union of the god Hermes with a shepardess. The god Helios was a child of the god Zeus and the mortal woman Euryphaessa. Helen of Troy was the child of Zeus and the woman Leda.

In 900 BC, the Aztec god Huitzilopochtli was conceived when his mother, Coatlicue, picked up a bunch of feathers and was impregnated by them. At the same time, the Aztec god Quetzalcoatl was born by a virgin named Chimalman who conceived when the god Onteol appeared to her in a dream. (The interesting thing is that this occurred in the Americas, which were unknown to people in Europe and the Middle East at the time. So these types of miraculous births were occurring around the world not just in the Middle East and Europe due to the state of human mind development at the time.)

In 771 BC, Romulus (Rome) was born from the union of the god Mars and the woman Aemila. In 600 BC, Zarathustra, a prophet in Persia was born to his mother, Dughdova, who was conceived by a shaft of light. Miraculous births occurred in China and Japan as well. In 563 BC, Maya fell asleep and was miraculously impregnated, resulting in the birth of Buddha.

So in those days, a few thousand years ago, gods were flying around everywhere. There were thousands of them. Priest, pharaohs, and kings were talking to them. Myths were abundant.

So the thinking at the time would naturally be that if Jesus were a god, there must have been some sort of miraculous or virgin birth.

Medical knowledge at the time was primitive. They didn't know about sperm, eggs, and fertilization. They probably saw a correlation between having sex and childbirth. But they also believed that illness and disease were caused by evil spirits and that fertility, pregnancy, and childbirth were overseen by the gods.

A story says that around 400 BC, Hippocrates, the father of modern medicine, testified in the case of a woman accused of adultery. The woman and her husband were white-skinned; her baby was black. The woman testified that the portrait of a Moor (black) hung over her bed, and in the height of passion with her husband, she focused on the picture of the Moor, and as a result, the baby was born with black skin. Hippocrates supported the possibility of her story, and she was vindicated.[14]

In 1637, in Grenoble, France, a wealthy woman, Madeleine d'Auvermont, gave birth to a boy despite the fact that her husband had been away for four years. She was tried for adultery. She testified that at night, in bed, she had thought intensely of her husband and conceived through the power of imagination. Physicians and theologians were consulted on the case, and they declared it was possible. She was found innocent.[14]

And even today, "miraculous" births occur. In 2013, the University of North Carolina reported on studying 7,870 adolescent and young adult females over a fourteen-year period. The report said that forty-five of the women in the group reported that they had been pregnant without having had sex.[15]

In January 2014, a nun in Italy gave birth to a baby boy, stating that she didn't even know she was pregnant.[14]

Was Jesus God?

Historian Bart Ehrman, in *How Jesus Became God,*[16] says that Jesus did proclaim to be the Messiah, but *Messiah* meant *leader* or *king,* not *God*. In those days, the Romans were the ones who decided who would be king. Jesus, a gifted storyteller, preacher, teacher, and leader, was attracting a growing following. This threatened the established aristocratic Jews who ran the temples. They brought it to the attention of the Romans who, sensing Jesus's growing power and desire to be king, eliminated him.

Ehrman says that in the four gospels of Matthew, Mark, Luke, and John, only in John does Jesus claim to be divine.

John was the last gospel written; if Jesus had claimed to be God, surely Matthew, Mark, and Luke would have reported that in their gospels (since it was kind of important). They didn't.

In fact, Mark 10:17 tells the story of the rich young man. "And as he (Jesus) was setting out on his journey, a man ran up and knelt before him and asked him, 'Good Teacher, what must I do to inherit eternal life?'

"And Jesus said to him, 'Why do you call me good? No one is good except God alone.'"

Papal Infallibility

On July 18, 1870, the Church ruled that all popes (including previous ones) were infallible when talking about Church doctrines and beliefs.[17]

At one time, popes were monarchs who controlled armies and broad swaths of land. Over the centuries, they lost most of this political power and control, and by 1870, all they had left was a large portion of land in central Italy, known as the Papal States, which still made the pope the largest landowner in the country.

But on September 20, 1870, the Italian Army took over all the land, leaving the Church with none. (In 1929, Mussolini granted the Church the land for Vatican City, creating a sovereign state.)[18]

So perhaps to overcome this loss of "political power," the Church created papal infallibility in order to be able to claim "spiritual power."

So wherever needed, the Church made up stuff like Mary's Immaculate Conception in 1854 and Mary's Assumption into Heaven in 1950 to try to retain some sense of influence and power.

Limbo

Augustine (354 AD–430 AD), a saint, said that infants who died without baptism, were consigned to Hell.

Later on, theologians made up Limbo, where these babies went and where they were deprived the vision of God (because they had not been baptized). This was certainly better than Hell.

In 2007, Pope Benedict XVI, after a three-year study, ruled there was no longer a Limbo.[19] He stated that they didn't have "sure knowledge" that there was no Limbo, but hoped that a "merciful God" would not deny "innocent babies entrance to Heaven." He did not say what happened to all the babies previously sent to Limbo over hundreds of years and did not explain how all the previous "infallible" Popes could have been wrong. Unbelievable.

Sanctity of Life

There is no such thing as "sanctity" of life; that's a made-up religious definition. Human life is special, but it's not sacred.

In fact, all life is special—dogs, cows, trees, plants, animals.

"Sacred" and "sanctity of life" have no place in government, where we supposedly have separation of church and state. Abortion is unfortunate; it is not a violation of the "sanctity of life."

Humans eat plants and animals because we're at the top of the food chain. We believe that there is nothing "morally" wrong with us doing so since we're the higher order of species.

But if aliens came here, making them a higher life form, would it be morally okay if they harvested us as food? We may not like it, but the answer is yes. If it's okay morally for us to eat cattle, then it's got to be okay for a higher life form to eat us. "Sanctity of life" has nothing to do with it. (Why is it morally all right to euthanize animals but not humans who are suffering at the end of their lives? This too will change.)

Tiny Sample

The virgin birth, Jesus as God, papal infallibility, Limbo, and sanctity of life is just a very tiny sample of religious "made-up stuff."

Chapter 26

Moses and the Commandments

Moses (1393 BC–1273 BC) was born in Egypt of Hebrew parents. The pharaoh at the time feared that a Hebrew leader would be born who would unite the Hebrew slaves against him. So he ordered that all Hebrew boys be drowned in the Nile. Moses's mother hid him near the riverbank, where he was discovered by the pharaoh's daughter, who adopted and named him.

Moses was raised in the pharaoh's household, where he had access to the latest knowledge. But after killing an Egyptian slave master, he fled across the Red Sea, where he encountered the God of Israel, who told him to return to Egypt and demand the release of the Israelites (Hebrews) held as slaves. Eventually, the pharaoh released the Hebrews, and Moses led them out of Egypt across the Red Sea to Mount Sinai, where he received the commandments from God. Then, after wandering in the desert for forty years, Moses led the Hebrews to the border of Israel, where he died at the age of 120.

Q: What Is The Bible?

A: The Bible Is The Written Word Of God . . . Its Principal

Author Is God, Though It Was Written By Men Whom God Enlightened And Moved To Write All Those Things, And Only Those Things, That He Wished To Be Written.

—*Baltimore Catechism*

The Bible—Old Testament

The Old Testament contains the Torah (Five Books of Moses), the Nevi'im (Book of Prophets), and the Ketuvim (Book of Writings).

God gave all of the teachings found in the Torah to Moses (except for the last eight verses of Deuteronomy, which describe Moses's death). The five books of the Torah—Genesis, Exodus, Leviticus, Numbers, and Deuteronomy—contain 613 commandments given to Moses by God at Mount Sinai, including the Ten Commandments. (The Jewish philosopher Maimonides (1135–1204) compiled this list of 613 commandments that are spread among the five books.)

Known collectively as the Law of Moses and the Word of God, these 613 commandments cover all aspects of life, including the worship of God, duties of priests, operations of temples, prayer, the poor, forbidden sexual practices, the judicial system, punishments, war, and, in general, how to live one's life.

Some of these 613 commandments[20, 21, 22, 23, 24] cover health practices (which may have been for religious purposes only or for healthful living).

A Few "Dietary" Commandments Out of Twenty-Six Total

- Not to eat unclean cattle.
- Not to eat unclean fish.
- Not to eat unclean fowl.

- Not to eat things that creep upon the earth.
- Not to eat any vermin of the earth.
- Not to eat animals that are found dead.
- Not to eat shellfish.
- Not to eat pork.

A Few "Agriculture" Commandments Out of Seven Total

- Not to crossbreed cattle of different species.
- Not to eat the fruit of a tree for three years from the time it was planted.

A Few "Purity" Commandments Out of Ten Total

- Foods become impure by contact with unclean things.
- Anyone who touches the carcass of an animal found dead will be unclean.
- A menstruating woman is unclean.

A Few "Leprosy" Commandments Out of Nine Total

- A leper is unclean.
- A leprous garment is unclean.
- A leprous house is unclean.
- The procedure of cleaning leprosy, whether of a man or of a house, takes place with cedar wood, hyssop, scarlet thread, two birds, and running water.

Illness, Disease, and Death In Moses's Time

As described in chapter 3, rampant illness, disease, and death (mentioned in numerous places in the Bible) were part of everyday life in ancient times.[25]

Smallpox (which has killed more people than any other disease in history), dysentery, jaundice, malaria, pneumonia, tuberculosis, polio, and typhoid fever struck and killed many. Upset stomach, intestinal troubles, headache, and arthritis were common. They had leprosy and occasionally, plague broke out. A minor infection could lead to death.[26, 27]

Farming techniques were primitive; famine occurred every few years, and people starved to death. Their limited diet lacked essential vitamins and minerals, causing malnutrition, which weakened the body, making it more susceptible to scurvy, anemia, and other illnesses that could result in death.[28]

While cleanliness was required of priests—for religious reasons, not for hygiene—it didn't filter down to the everyday people. They lived in unsanitary conditions. Since they had no plumbing, water was used sparingly and probably never for cleaning their homes or belongings. The dirt floors invited insects and rodents, spreading bacteria and disease. Their drinking water contained harmful parasites, causing illness and death. Human and other wastes were dumped in fields or canals or the river where they bathed.[29, 30]

Thirty percent of newborns died at childbirth, along with a great many mothers. Once weaned from the mother, a young child could be infected by bacteria and parasites in water and food, causing diarrhea that could result in death.[31]

Their "medicine"—dung from various animals, fly droppings, lizard blood, etc.—could kill a person.

People died because infected teeth weren't pulled.

Most people were dead by age forty.

Mortality Rates	**Ancient Egypt**	**Today's Egypt**	**Today's US**
Infant Mortality Rate: First Year	30%	4%	½%
Childhood Mortality Rate: Ages 1–5	20%	5%	2%
Life Expectancy at Birth[32]	20–30 yrs	70 yrs	80 yrs

They believed that illness and disease came from God(s) and evil spirits. As a result, they thought that the cure was prayer and magic. Religion ruled science and medicine.

If You Were Moses's God

So, if you were Moses's God, would you have been more interested in being praised and worshipped or in helping the people that you created?

Moses's God

If you read each of Moses's 613 commandments, you quickly realize that each commandment would have been recognized and understood as logical, commonsense advice by most people of the day. But they contained no new knowledge that you might expect from a God.

The number one challenge to the people of the time was staying alive. Moses's God issued 613 commandments, a few of which contained some very limited, basic, primitive instructions on cleanliness and hygiene, what animals to eat and not eat, agricultural practices, and the disease of leprosy.

Why didn't Moses's God command the people:

- to boil drinking water to kill the parasites;
- to not bathe in dirty or polluted waters;
- to keep insects and rodents out of their houses;
- to throw old food out to minimize bacteria;
- to store food properly to minimize illness;
- not to dispose of waste in rivers or canals;
- to avoid and reduce pools of water containing mosquitoes;
- in better farming techniques to eliminate famine and starvation;

- to eat foods with proper vitamins and minerals to reduce scurvy and anemia;
- on what to eat for a healthy diet (there were commandments on what not to eat);
- on how to reduce infant and childhood mortality;
- to extract infected teeth;
- to not use animal dung for "medicine";
- that some illnesses and disease are contagious;
- to prevent and minimize illness and disease?

Leprosy & Leviticus

Leprosy is an easily visible disease that causes disfiguration of the skin, muscle weakness, and nerve damage. Caused by bacteria, it can take from three to twenty years after contact with the bacteria for symptoms to appear.

People at highest risk are those living near polluted water, with a poor diet and a weakened immune system. It can be spread by respiratory droplets, but it's not very contagious. It can eventually lead to death by secondary infections. Fortunately, 95 percent of people are immune to the bacteria.

Other diseases, such as smallpox, tuberculosis, malaria, dysentery, typhoid fever, jaundice, pneumonia, polio, cholera, plague, and famine are far deadlier and killed many, many millions more people in ancient times.

When it came to illness and disease, leprosy was the least of their worries.

But in the Bible, in the Book of Leviticus (written by Moses) more than two thousand words are dedicated to dealing with leprosy—little to none to the other diseases. How could Moses's God spend so much attention on leprosy and virtually none on the other diseases that killed many millions more people? Would a "real" God talk only about leprosy and ignore these other far more deadly diseases? Obviously, this doesn't make sense.

The Truth about Moses's God

Why was Moses's God (who liked to issue commandments) so critically lacking in these crucial areas?

And why didn't Moses's God know that menstruating women are not unclean, that pork properly cooked is safe to eat, and that shellfish—harvested during the right season and not from polluted waters—properly cooked is safe to eat?

The answer to these questions is simple, and, . . . the answer is obvious.

The reason Moses's God never addressed these questions is because Moses never talked to a "real God." He either made up or dreamed up his "God."

Moses and the other Bible writers had no idea that proper sanitation and clean water could reduce and prevent illness and disease. They didn't know that bacteria, viruses, and fungi could cause disease. They thought that illness and disease came from God, and the only possible cure was prayer to God.

If there were a "real God," He, She, or It would know all of these things and more and would know how to reduce, prevent, and eliminate illness and disease. While Moses's God gave a few primitive commandments (which all people of the time were already aware of) regarding cleanliness, sanitation, etc., he could have given far many more "advanced" instructions in these areas. This alone proves that Moses never talked to a "real God."

So Moses's God was a made-up god just like all the other thousands of gods that "existed" at that time. Moses used the concept of "one god" to help organize, control, and unite the people of his time for the common good; people who worshipped thousands of gods were probably difficult to organize and control. To make matters worse, the "God" that Moses invented was rather inferior, insecure, violent, and not very helpful; a "God" of very few standards, a "God" that we would not admire or respect today. (Where is this "God" today, when we could use a cure for cancer? Nowhere to be found. This "God" didn't even have the foresight to realize that hundreds of his 613 commandments would become quickly out of date; he had no mechanism for updating the commandments.)

The lack of advice and instruction concerning hygiene, sanitation, and illness prevention exposed Moses's God as a made-up God, and proves that the Bible is not the "word of God" but the "word of men."

Chapter 27

Birth, Marriage, Death, & Intuition

Years ago, most of the people I knew went to church weekly and more often during Lent. Now my sense is that most of them only go the church for baptisms, weddings, and funerals.

In 2014, the Gallup Poll said that weekly church attendance in the 1950s and early 1960s was 50 percent of all people in the United States, and that in those days, 70 percent of the people said that religion was very important to them.[33, 34] Today, Gallup says that 40 percent of people attend weekly church service.[35] But the 40 percent attendance number is disputed. Kirk Hadaway and Penny Marler in the 2005 *Journal for the Scientific Study of Religion* say that only 20 percent of people actually go to church weekly.[36]

My own, totally unscientific, drive-by church parking-lot survey tells me that the number feels like 10 percent or even less. (Actually, I think that there are more cars in gym parking lots on Sunday. Today, stores are open on Sunday, youth sports are ubiquitous, and there are many other things to do, unlike a few decades ago. Did we go to Church in those days because there was nothing else to do?)

The GFK/AP survey of March 2014 asked people how often they attended religious services outside of weddings and funerals. The responses were:

Never	25%
Less than a few times per year	18%
A few times per year	12%
Once or twice a month	11%
Once a week	20%
More than once a week	12%
No answer	2%

The Center for Applied Research in the Apostate (CARA)[37] at Georgetown University in 2014 published the following statistics:

United States

	1965	**2014**
Total priests	58,632	38,275
Total nuns	179,954	49,883
Students in Catholic Elementary Schools	4,430,000	1,380,000
Baptisms	1,310,000	773,302
Marriages	352,458	154,450
Weekly Mass attendance	55%	24%
Total number of Catholics	46,300,000	66,600,000

World

	1965	**2014**
Total priests	419,728	414,313
Total number of Catholics	654,000,000	1,230,000,000
% of world population	18%	17%

So, in the United States, the numbers of priests, nuns, students in Catholic elementary school, baptisms, marriages, and weekly Mass attendance are down dramatically while the total number of people calling themselves Catholics has grown by 50 percent.

In the world, the total number of priests has stayed the same while the total number of Catholics has doubled reflecting the overall growth of the world population during this time.

What's going on?

Researchers say that the decline in church attendance is due to many factors, such as:

- Boredom during services
- Lack of motivation
- Incompatibility of some church beliefs
- Other priorities/too busy
- Changing demographics
- Poor Church leadership.[38, 39]

In 2010, the University of Chicago's General Social Survey said that the number of people who claim no religious affiliations grew from 8 percent in 1990 to 18 percent in 2010, an increase of twenty-five million people.

The Church is growing in third-world countries, where people deal with hunger, disease, oppression, and a general lack of knowledge. In advanced countries like the United States, the gradual spread of knowledge correlates with the decline in Church attendance. As we gain knowledge, do we somehow realize that gradually, subconsciously religion doesn't make sense?

Weddings and funerals require a building where people can gather. Is that why we still go to Church on those occasions? Because of the convenient building?

Most people would say that the most special, most interesting, most enjoyable parts of weddings and funerals are not the church services, or the words of a priest, minister, or rabbi, but rather, the gathering of friends, the speeches, the eulogies, the joy, the laughter, the tears.

If we truly believed in Heaven, religion, and God, wouldn't they be the most important things in our lives? Wouldn't we all be trying to indirectly get into life-ending "accidents" to go to Heaven immediately? And if the dead were truly going to Heaven, even if they're a child or younger person dying before their times, shouldn't we rejoice and be happy? Why the tears? Is it because intuitively, subconsciously we "know" that there is no Heaven, that there is no God?

Chapter 28

If

If you saw me standing on top of a building, would you believe that I flew up there? Would that be proof that I could fly?

If your team wins a big game, does that prove that God is on your side?

If I get food poisoning from shellfish or pork, does that prove that God is punishing me?

If you or a loved one recovers from a serious illness, does that prove that prayer works?

If an earthquake strikes California, does that prove that God is angry with Hollywood?

If a hurricane strikes New Orleans, does that mean that God is punishing them for sinful practices?

If a meteor crashes into Siberia, does that prove that God is angry with Russia?

If we have a severe thunderstorm, does that mean that Zeus or Thor caused it?

If a magician saws a lady lying in a box in half and then puts her back together, is that proof of a miracle?

If people, a long time ago, believed that the earth was flat, did that prove that the earth was flat?

If you see the sunrise in the morning and set in the evening, does that prove that the sun circles around the earth?

If a heretic confessed under torture, does that prove that he was guilty?

If when you awoke on Christmas morning as a child and saw a bicycle near the Christmas tree, did that prove that Santa's elves made it and Santa delivered it?

If a preacher—a man of God—tells you that God spoke to him the previous night, does that prove that God spoke to him?

If a great multitude of gifted artists over many centuries created wondrous artwork—paintings, statues, cathedrals, music—depicting God, does that prove that God exists? (A great many pagan, Egyptian, Greek, Roman, and other ancient artists also created wonderful artwork of their gods.)

If a lot of decent people believe in God, does that prove that God exists?

If you marvel at the simplicity, complexity, and diversity of trees and animals, humans, plants, and stars, does that prove that God made them?

A few thousand years ago, most people would probably have answered yes to most or all of these questions. And if we lived then, we would have answered yes as well. But a tremendous amount of knowledge has developed over the last few thousand years to the point that, today, most of us would answer no to these questions. (Shockingly, today, in the Third World, and in this country as well, there are far too many people who would still answer yes.)

Since we have no proof that God exists, we must conclude that there is no God.

Chapter 29

The Greatest Ponzi Scheme of All Time

Ponzi Schemes

Charles Ponzi, a charming, impeccably dressed, flamboyant con man, was born in Italy in 1882.[40,41] In 1907, he moved to Canada and spent three years in prison for forgery. He moved to the United States in 1911 and spent two years in prison for smuggling people across the border.

Then he moved to Boston and thought he saw an opportunity to make money based on the difference in postage rates from country to country. A person in Italy could send a letter to a friend in the United States and include in that letter an “international reply coupon,” which could be redeemed by the US friend for free US postage to send a letter back to Italy. Since postage rates in Italy were slightly lower than in the United States, the international reply coupon (bought for a lower price in Italy) could be exchanged for a higher-priced US stamp.

Ponzi figured that if he bought enough of these lower cost international reply coupons in Italy and exchanged them for US stamps, he could then sell the US stamps and make a profit. While this was legal, the profit margin (difference in price of US and Italian stamps) was so small that after administrative costs, there was no profit left.

Nevertheless, Ponzi understood that the buying and selling of international reply coupons would sound mysterious, intriguing, and exciting to gullible investors. So in early 1920, at the age of thirty-eight, he set up the Securities Exchange Company.

He told investors that through his manipulation of international reply coupons he could double their money in ninety days. Word spread; investors came. As promised, he paid off some of the early investors much quicker than ninety days (with the money he took in from later investors). Word continued to spread; now this was spiraling upward. Early investors didn't want to take their "profits" out; they wanted Ponzi to keep reinvesting their "imagined gains." The press did positive stories and more money flowed in; when the first negative stories appeared, even more money flowed in.

Ponzi bought a mansion and lived a lavish life, but eventually further investigations and news stories led to a collapse in August 1920. In all, in just eight months, the investors lost $20 million (equal to $225 million today) and six banks collapsed.

Ponzi pleaded guilty to federal charges and served three and a half years in prison. He was also convicted by the state of Massachusetts and sentenced to nine years in prison. While appealing his Massachusetts sentence, he jumped bail and fled to Florida, where, in 1925, he was found guilty of a scam to sell swampland. He was eventually sent back to Massachusetts to serve out his sentence. In 1934, he was deported to Italy, and after more schemes, made his way to Brazil, where he died poor at the age of sixty-seven.

But Ponzi had generated so much national attention that, even today, almost a hundred years later, scams in which early investors are repaid not from true gains in investments but from funds invested by later investors (thereby attracting even more investors) are called Ponzi schemes. So, in a Ponzi scheme, people invest their money into something that doesn't exist, fueled by the enthusiasm of others who have previously invested.

> "Even if they never got anything for it, it was cheap at that price. I had given them the best show that was ever staged in their territory since the landing of the Pilgrims."
>
> —Charles Ponzi (1882–1949), con artist

Sarah Howe

In 1880, Sarah Howe opened the Ladies Deposit Company in Boston.[42] It was a specialty investment firm only for women. She promised the women an 8 percent per month interest on their deposits and to "help" the women maintain their investment, they could only withdraw interest, never the principal. She used the principal to pay interest and took the remainder for herself. One thousand two hundred women lost today's equivalent of $11 million; Sarah spent three years in prison.

J. David Dominelli

In 1979, David Dominelli opened J. David & Company, an investment firm in La Jolla, California. Known as "Captain Money," Dominelli promised high-profile wealthy investors a 40 percent to 50 percent gain on their money. One thousand five hundred investors lost $80 million; Dominelli spent ten years in prison.[43]

Gerald Payne

Gerald Payne, a Florida contractor, found Jesus in the 1980s and became a minister. In 1993, Payne founded Greater Ministries International, which promised conservative Christian groups that they could double their money through his investments in precious metals. Citing Luke 6:38, "Give, and it shall be given unto you," Payne managed to bilk 18,000 believers out of $500 million in the name of God. He was sentenced to twenty-seven years in prison.[44]

> "God said this (investment) plan will be around till He comes again. Praise God."
>
> —Gerald Payne (1936–), preacher

Allen Stanford

Allen Stanford grew up in Texas and together with his father made a lot of money in real estate in the early 1980s. Then, he set up Stanford International Bank in Antigua in the mid-1980s, promising investors higher than market returns. He owned businesses, real estate, and newspapers in Antigua and became so well known and respected that he was made a knight with the title "Sir Allen" in 2006. He enjoyed luxury and cricket.

He was arrested in 2009 for running a massive Ponzi scheme at the bank, was convicted in 2012, and is serving a sentence of 110 years. Thousands of people lost a total of $7 billion.[45]

Bernie Madoff

Bernie Madoff's grandparents emigrated from Poland; his mom was a homemaker, his father a plumber in Queens, New York. Born in 1938, he worked hard and opened Bernard L. Madoff Securities Company in 1960. The company traded stocks on the NASDAQ using innovative computer technology. He was a philanthropist donating to many charities. He lived in Manhattan's Upper East Side and had homes in the Hamptons, France, and Palm Beach.

Around 1990, he began an investment scam promising investors a 10.5 percent annual return, and ran it for twenty years. Investment money poured in, which he deposited in his personal bank account. He took advantage of banks, charities, and thousands of individuals, many of whom were well-known, wealthy celebrities. He was arrested in 2008, pled guilty in 2009, and was sentenced to 150 years in prison. Investors lost $10 billion to $20 billion, making it the largest Ponzi scheme ever.[46]

These are just a few of thousands of these types of fraudulent schemes. There are many other legal and illegal swindles, such as some multilevel marketing companies, casinos, lotteries, and work-at-home scams, to mention a few. They all take advantage of trusting people and are masters at manipulating those people's hopes and dreams.

Some Religious Scams

Faith Healers

Faith healers say that they can heal the body of illnesses and disease through their "connection with Jesus/God" without medical intervention.[47] This fraud is nothing new; this is what the priests told the people in ancient Egypt.

In 1955, Faith Healer Jack Coe told the parents of a three-year-old boy that he had cured him of polio. He told the parents to remove the boy's leg braces, causing him considerable pain.

The boy, of course, was not cured. Ironically, Coe was diagnosed with polio shortly thereafter and died in 1956 at the age of thirty-eight—he couldn't save himself.

In 1967, Dr. William A. Nolen[48] conducted a follow-up study on twenty-three people "healed" by Faith Healer Kathryn Kuhlman.[49] His research showed that there were no cures of the twenty-three. Kuhlman "cured" a woman of spinal cancer and instructed her to throw away her back brace and run across the stage. The next day the woman's spine collapsed, and she died four months later. Kuhlman had open-heart surgery in 1976 and died at age sixty-nine. She couldn't heal herself. A plaque "honoring" her is in the city park of Concordia, Missouri.

In 1998, the American Cancer Society[50] investigated 172 cases of death of children who were treated by faith healing without any conventional medicine. They determined that 90 percent of the children would have lived if given proper medical treatment, and the other 10 percent would have had a fighting chance of survival.

At Benny Hinn's[51, 52] "Miracle Crusade" meetings, where he claims that God communicates with and works through him, people with serious medical conditions—quadriplegia, brain damage, cerebral palsy, and other visible physical conditions—are not allowed on stage to be "cured."[53, 54] Only those with conditions that cannot be seen are allowed to be "cured."

Shouldn't "faith healers," like Pat Robertson, Peter Popoff, Kenneth Copeland, Richard Rossi, and Billy Burke be healing war veterans who have lost limbs? Shouldn't their "connection" with God allow them to get these limbs fully restored? Which of these "faith healers" has fully restored an arm or a leg?

"Faith healers" take advantage of those less fortunate, they lift their hopes up, and then, after the "healers" leave town, the victims come crashing down as they realize they have not been healed. On which level of Hell would Dante relegate “faith healers”?

Prosperity Preachers

These preachers tell the hopeful (who are usually less well off) that God wants them to be rich, and if they have faith and make donations to the preachers, wealth will surely come their way.[55] They distort Bible passages, telling the gullible that God (with the prodding of the preachers who are connected to God), will shower them with riches. They neglect to quote Matthew 14:14, in which Jesus said, “It is easier for a camel to go through the eye of a needle than for someone who is rich to enter the Kingdom of God.”

Preachers such as E. W. Kenyon, Oral Roberts, A. A. Allen, Creflo Dollar, Kenneth Hagen, Edir Macedo, Joel Osteen, and Jim and Tammy Faye Bakker had or have large followings, TV programs, best-selling books, expensive cars, big houses, and large bank accounts.

These are all Ponzi Schemes—the believers donate their money and receive nothing in return.

Why don’t regulators shut them down? In a financial Ponzi scheme, people “invest” their money hoping for high returns; with faith healers and prosperity preachers, people “donate” (give away) their money, hoping for cures and wealth.

How many of these preachers would willingly submit to a polygraph exam inquiring whether they truly believed in God, or were just in it for the money?

Mother Teresa

Whether you’re religious or not, you have to admire Mother Teresa, who, despite her faults, dedicated her life to help the poor and sick in India.[56] The religious order she founded ran hospices: took care of people with AIDS, leprosy, and tuberculosis; and ran soup kitchens, clinics, orphanages, and schools. Members of her order took vows of chastity, poverty, and service to the poorest of the poor. Which of the previous preachers would do that? Which ones would measure up to Mother Teresa?

The problem isn't that con men exist; con men have always existed and will always exist. The problem is that desperate people, overcome with grief and hoping for a better life, lack the individual strength and courage to face life as it is, making them easy prey.

In general, those who talk the most about God and Jesus are the ones to trust the least.

Other Foolishness

Snake Handling

George Went Hensley,[57] a Pentecostal minister, believed that the Bible required all Christians to handle snakes. Starting in 1910, he conducted snake-handling religious services. Other preachers took up the practice; occasionally people died. Hensley himself died of a snakebite in 1955.

IHOP

Not pancakes, but the International House of Prayer,[58] which was founded in 1999, was reported in a 2013 PBS documentary to be partly responsible for Uganda's Anti-Homosexuality Law. The law initially included the death penalty for homosexuality; a later revision changed it to life in prison. It is currently under review.

Blood Moons

Preachers John Hagee and Mark Blitz have predicted a coming series of lunar eclipses portend the end of days. A reddish color on a full moon (blood moon) during a lunar eclipse is commonplace.[59,60] (The reddish color is caused by the sun's light traveling through the earth's atmosphere, as it does during pretty sunrises and sunsets.) They've even published a book and, amazingly, some people are buying it.

God Bless America

There's no shortage of phony politicians who invoke the slogan "God Bless America" to pander to the religious right. (Should presidential candidates be disqualified if they fervently believe in God, since it demonstrates that they lack curiosity, brainpower, and honesty?)

How to Judge a Preacher

Many books have been written about thousands of fraudulent, lunatic preachers. So, how do you judge a preacher? Very simply—by their bank accounts—the more money, the bigger the liar.

Jesus and the Vatican

In 2004, in order to help pay for clergy sexual abuse settlements, the Archbishop of Boston's mansion and grounds were sold for $99 million. The Cardinal of New York City lives in a 15,000-square-foot mansion valued at $30 million. The Cardinal of Chicago lives in a mansion worth $14 million. A former German Archbishop has built a mansion worth $42 million. According to CNN, in the United States, ten of thirty-four Catholic Archbishops live in dwellings valued at more than $1 million.[61]

The Vatican, a sovereign state with diplomatic immunity (giving it some protection from sexual abuse cases and other crimes), is said to be one of the world's largest landowners. Its wealth includes real estate, art collections, ancient books and manuscripts, jewels, gold, stocks, bonds, cash, and investments.

If Jesus visited us today, wouldn't he tell the Vatican to sell its material possessions and distribute the proceeds to the poor?

(When a snowstorm closes a restaurant, they never make up lost revenue—people don't come in the next day and order two meals. If you skip Mass on Sunday, you don't double up your donation the next Sunday. Is that why it's a sin to miss Mass?)

Summing Up

So, let me see if I understand this: I give you (organized religion) a never-ending stream of money and control over my life, and you promise me that when I die (and have no recourse if you don't fulfill your end of the deal) that you will get me into Heaven. Absolutely, insanely unbelievable.

Charles Ponzi and Bernie Madoff were not even close to being in organized religion's league when it came to scams—they were pikers.

The Greatest Ponzi Scheme of all time is . . . organized religion. (And on top of it, they're tax-free.)

Some Thoughts by Some Preachers

"God is a being that stands somewhere around six-foot-two or six-foot-three, weighs somewhere in the neighborhood of a couple hundred pounds, and has a hand span of nine inches across."[62]

—Kenneth Copeland (1936–), preacher

"When you don't give money, it shows that you have the devil's nature."[63]

—Benny Hinn (1952–), faith healer

"I believe that Hurricane Katrina was, in fact, the judgment of God against the city of New Orleans . . . New Orleans had a level of sin that was offensive to God."[64]

—John Hagee (1940–), preacher

"With all due respect to those dear people . . . God does not hear the prayer of a Jew."[65]

—Bailey Smith (1957–), preacher

"Evolution is a bankrupt speculative philosophy, not a scientific fact."[66]

—Jimmy Swaggart (1935–), preacher

"God will not move unless I say it. Why? Because He made us coworkers with Him. He set things up that way."[67]

—Benny Hinn (1952–), faith healer

"When you come to Heritage USA, remember to bring your Bible and your Visa card . . . God doesn't take American Express."[68]

—Jim Bakker (1940–), preacher

"Many of those people involved with Hitler were Satanists. Many were homosexuals. The two things seem to go together."[69]

—Pat Robertson (1930–), preacher

"I believe that homosexuality is shown as a sin in the Scripture."[70]

—Joel Osteen (1963–), prosperity preacher

"If you have attacked me, your children will pay for it."[63]

—Benny Hinn (1952–), faith healer

"AIDS is not just God's punishment for homosexuals; it is God's punishment for the society that tolerates homosexuals."[71]

—Jerry Falwell (1933–2007), preacher

"If I do not return to the pulpit this weekend, millions of people will go to Hell."[66]

—Jimmy Swaggart (1935–), preacher

"You want to prosper? Money will be falling on you, from left, right, and center. God will begin to prosper you, for money always follows righteousness."[67]

—Benny Hinn (1952–), faith healer

"Christ is the head of the household, and the husband is the head of the wife, and that's the way it is, period."[72]

—Pat Robertson (1930–), preacher

"I believe that the heavens are God's billboard, that he has been sending signals to planet Earth. God is literally screaming at the world, 'I'm coming soon.'"[73]

—John Hagee (1940–), preacher

"I take Jesus shopping with me. I say, 'OK, Jesus, help me find a bargain.'"[74]

—Tammy Faye Bakker (1942–2007), preacher

"As you use your faith, God is going to wipe out your credit card debts."[75]

—Mike Murdock (1946–), prosperity preacher

"God said, 'It is time to tell money you don't belong to the wicked, you belong to us. Money come to me now.'"[75]

—Leroy Thompson (1944–), prosperity preacher

"We look at gay being . . . a bigger sin than being proud . . . or not telling the truth."[76]

—Joel Osteen (1963–), prosperity preacher

"Homosexuality means the death of society because homosexuals can recruit, but they cannot reproduce."[77]

—John Hagee (1940–), preacher

"Feminism encourages women to leave their husbands, kill their children, practice witchcraft, destroy capitalism, and become lesbians."[69]

—Pat Robertson (1930–), preacher

"Poverty is from the devil and God wants all Christians to be prosperous."[67]

—Benny Hinn (1952–), faith healer

"I bought some crystal meth, but I threw it away."[78]

—Ted Haggard (1956–), preacher

"I gave my heart to the Lord, and I remember the incident vividly. The Lord spoke to me."[66]

—Jimmy Swaggart (1935–), preacher

"Sometimes I wish God would give me a Holy Ghost machine gun. I'd blow your head off."[75]

—Benny Hinn (1952–), faith healer

"Ebola is God's judgment on America for trying to divide Jerusalem."[79]

—John Hagee (1940–), preacher

"When students are taught that they are no different from animals, they act like it."[80]

—Rick Warren (1954–), preacher

"Good Christians, like slaves and soldiers, ask no questions."[71]

—Jerry Falwell (1933–2007), preacher

"I heard Satan say, 'Jesus is playing you for a sucker.'"[69]

—Pat Robertson (1930–), preacher

"I saw Jesus walk into my bedroom."[67]

—Benny Hinn (1952–), faith healer

"The flooding of New Orleans is a sign that God is tired of seeing his creation mocked by the Mardi Gras and its perverted display of debauchery and exposed breasts."[69]

—Pat Robertson (1930–), preacher

"I don't believe anyone begins a homosexual."[71]

—Jerry Falwell (1933–2007), preacher

Said in 1989: "The Lord tells me that in the mid-1990s God will destroy the homosexual community of America."[63]

—Benny Hinn (1952–), faith healer

Now, Some Intelligent Thoughts

"Of all bad men, religious bad men are the worst."

—C. S. Lewis (1898–1963), author

"I distrust those people who know so well what God wants them to do because I notice it always coincides with their own desires."

—Susan B. Anthony (1820–1906), social reformer

"I think (God) would prefer an honest and righteous atheist to a TV preacher whose every word is 'God, God, God' and whose every deed is 'foul, foul, foul.'"

—Isaac Asimov (1920–1992), author

"The clergy converted the simple teachings of Jesus into an engine for enslaving mankind and . . . to filch wealth and power to themselves . . . these clergy, in fact, constitute the real anti-Christ."

—Thomas Jefferson (1743–1826), third president

"He loves you, and he needs money! He's all-powerful, all-perfect, all-knowing, all-wise, and somehow just can't handle money."

—George Carlin (1937–2008), comedian

"It's an incredible con job . . . to believe something now in exchange for something after death. Even corporations with their reward systems don't try to make it posthumous."

—Gloria Steinem (1934–), journalist

"You can fool some of the people some of the time—and that's enough to make a decent living."

—W. C. Fields (1880–1946), comedian

Chapter 30

Atheists and Agnostics

Atheists believe that there is no proof of God; agnostics aren't quite sure but they lean that way.

The word *atheist* has a harsh, negative feel almost like, *murderer, kidnapper,* and *rapist*. *Agnostic* sounds puzzling. We need a more "comfortable" term to describe people who don't believe in God. *Realist, reasoner, faithless, doubter, skeptic, unreligious, nonreligious, unbeliever, misbeliever,* and *disbeliever* are some possibilities.

What do we call people who believe in God? We call them faithful, religious, and believers.

What do we call people who believe in Santa? Do we call them *Santa-believers*? No. We call them children. And those who don't believe in Santa are not called non-Santa-believers. They're called teenagers and adults.

Since we have no proof that God exists (as with Santa), nonbelief is the normal state. But society isn't quite ready to accept "nonbelief" as the normal state, so those who believe, let's call them believers, and to take the harshness out of the word *atheist*, those who don't could be called nonbelievers.

Chapter 31

You're Not Alone

As you begin to doubt religion and God, you may feel like you're alone; you're not.

A 2012 Pew Research Center study said that the number of Americans who do not identify with any religion is growing at a rapid pace. Gallup in a 2008 poll said that one-third of Americans say that religion is not important in their daily life. Pew also said that 30 percent of the respondents were agnostics or atheists.

Most people do not publicize their religious beliefs, some do.

According to Wikipedia, here are a few well-known atheists and agnostics—nonbelievers. (There are, of course, thousands of additional "famous" people who don't publicize that they're nonbelievers.)

A Few Well-Known Nonbelievers[81, 82]

Woody Allen, film director
Robert Altman, film director
Kingsley Amis, author
Philip Anderson, scientist
Susan B. Anthony, activist
Lance Armstrong, cyclist
Jane Asher, actress
Isaac Asimov, author
Clement Atlee, politician
David Attenborough, anthropologist
Francis Bacon, artist

Kevin Bacon, actor
James Baldwin, actor
Javier Bardem, actor
Samuel Beckett, playwright
Alexander Graham Bell, scientist
David Ben-Gurion, politician
Ingmar Bergman, film director
Irving Berlin, composer
Georges Bizet, composer
Niels Bohr, physicist
Johannes Brahms, composer
Marlon Brando, actor
Richard Branson, businessperson
Bertolt Brecht, poet
Jacob Bronowski, biologist
Warren Buffet, businessperson
Richard Burton, actor
Gabriel Byrne, actor
James Cameron, film director
Albert Camus, philosopher
George Carlin, comedian
Thomas Carlyle, author
Andrew Carnegie, businessperson
Dick Cavett, TV host
Charlie Chaplin, actor
Anton Chekhov, author
Noam Chomsky, activist
Winston Churchill, politician
Arthur C. Clark, author
Georges Clemenceau, politician
Joseph Conrad, author
Aaron Copland, composer
Noel Coward, actor
Francis Crick, scientist
Marie Curie, scientist
Salvador Dali, artist
Rodney Dangerfield, comedian
Clarence Darrow, lawyer
Larry David, comedian
Richard Dawkins, biologist
Moshe Dayan, politician
Simone de Beauvoir, philosopher
Neil de Grasse Tyson, physicist
Marlene Dietrich, actress
Phyllis Diller, comedian
Stanley Donen, film director
Arthur Conan Doyle, author
Richard Dreyfuss, actor
Roger Ebert, film critic
Zac Efron, actor
Albert Einstein, physicist
George Eliot, author
David Feherty, golfer
Enrico Fermi, physicist
Richard Feynman, physicist
Carrie Fisher, actress
Ken Follett, author
Henry Fonda, actor
Peter Fonda, actor
E. M. Forster, author
Jodie Foster, actress
John Fowles, author
Betty Friedan, activist
Sigmund Freud, psychologist
David Friedman, writer
Milton Friedman, economist
Erich Fromm, psychologist
Stephen Fry, author
Giuseppe Garibaldi, politician
Bill Gates, businessperson
Ricky Gervais, comedian
Paul Giamatti, actor
Mikhail Gorbachev, politician
Stephen Jay Gould, biologist

Graham Greene, author
Kathy Griffin, comedian
Alan Guth, physicist
Alan Hale, astronomer
J. B. S. Haldane, biologist
Edmond Halley, astronomer
Thomas Hardy, author
Neil Patrick Harris, actor
Stephen Hawking, physicist
Joseph Heller, author
Ernest Hemingway, author
Katherine Hepburn, actress
Peter Higgs, physicist
Edmund Hillary, mountaineer
Christopher Hitchens, author
Abbie Hoffman, activist
Francois Hollande, politician
A. E. Housman, poet
Fred Hoyle, astronomer
Edwin Hubble, astronomer
John Huston, film director
Aldous Huxley, author
Henrik Ibsen, playwright
Penn Jillette, illusionist
Billy Joel, musician
James Joyce, author
Franz Kafka, writer
Immanuel Kant, philosopher
John Keats, poet
Gene Kelly, dancer
John Maynard Keynes, economist
Omar Khayyam, poet
Margot Kidder, actress
Larry King, TV host
Alfred Kinsey, biologist
Michael Kinsley, journalist
Keira Knightley, actress
Laurence Krauss, physicist
Stanley Kubrick, film director
Ray Kurzweil, scientist
Burt Lancaster, actor
Christopher Langton, scientist
Charles Laughton, actor
Cloris Leachman, actress
Richard Leakey, archeologist
Bruce Lee, actor
John Lennon, musician
Jack London, author
Percival Lowell, astronomer
Seth MacFarlane, comedian
Bill Maher, comedian
John Malkovich, actor
Thomas Mann, author
Abraham Maslow, psychologist
Henri Matisse, artist
Dave Matthews, musician
W. Somerset Maugham, author
Paul Mazursky, film director
Mary McCarthy, author
Ian McKellen, actor
Butterfly McQueen, actress
Golda Meir, politician
H. L. Mencken, journalist
John Stuart Mill, philosopher
Arthur Miller, playwright
Helen Mirren, actress
Claude Monet, artist
Julianne Moore, actress
Desmond Morris, zoologist
Elon Musk, businessperson
Rafal Nadal, tennis player
Jawaharlal Nehru, politician
Mike Nichols, film director
Jack Nicholson, actor

Friedrich Nietzsche, philosopher
Alfred Nobel, chemist
Bill Nye, scientist
Diana Nyad, swimmer
Frank Oppenheimer, physicist
George Orwell, author
Eugene O'Neill, playwright
Joyce Carol Oates, author
Barack Obama, Sr., economist
Charlie Parker, musician
Linus Pauling, chemist
Ivan Pavlov, scientist
Sean Penn, actor
Saul Perlmutter, physicist
Joaquin Phoenix, actor
Pablo Picasso, artist
Alan Pinkerton, detective
Brad Pitt, actor
Henri Poincare, mathematician
Simeon Denis Poisson, mathematician
Sidney Poitier, actor
Marcel Proust, author
Alexander Pushkin, author
Yitzhak Rabin, politician
Daniel Radcliffe, actor
Ayn Rand, author
Maurice Ravel, composer
Keanu Reeves, actor
Carl Reiner, comedian
Rob Reiner, film director
Gene Roddenbury, author
Richard Rodgers, composer
Ray Romano, actor
Linda Ronstadt, musician
Andy Rooney, journalist
Philip Roth, author
Bertrand Russell, philosopher
Carl Sagan, astronomer
Andre Sakharov, scientist
Margaret Sanger, activist
George Santayana, philosopher
Robert Sapolsky, scientist
Jean-Paul Sartre, philosopher
Arthur Schlesinger, historian
Erwin Schrodinger, physicist
Franz Shubert, composer
Robert Schumann, composer
George C. Scott, actor
Ridley Scott, film director
George Bernard Shaw, author
Mary Shelley, author
Percy Bysshe Shelley, author
William Shockley, physicist
Dmitri Shostakovich, composer
Clive Sinclair, scientist
B. F. Skinner, psychologist
George Smoot, physicist
Steven Soderbergh, filmmaker
Annika Sorenstam, tennis player
George Soros, businessperson
Benedict Spinoza, philosopher
John Steinbeck, author
Howard Stern, radio host
Robert Louis Stevenson, author
Sting, musician
Leonard Susskind, scientist
Quentin Tarantino, film director
Edward Teller, physicist
Teller, illusionist
Emma Thompson, actress
Pat Tillman, athlete/soldier
Francois Truffaut, filmmaker
Alan Turing, mathematician

Ted Turner, businessperson
Mark Twain, author
Vincent Van Gogh, artist
Giuseppe Verdi, composer
Jesse Ventura, wrestler
Gore Vidal, author
Kurt Vonnegut, author
John von Newmann, mathematician
James D. Watson, scientist
H. G. Wells, author
Stephen Weinberg, scientist
Elie Wiesel, author
Gene Wilder, actor
Vaughan Williams, composer
Tom Wolfe, author
Virginia Woolf, author
Steve Wozniak, businessperson
Frank Lloyd Wright, architect
Émile Zola, author
Mark Zuckerberg, businessperson

Chapter 32

"Love Thy Neighbor"

If the real message of Jesus was to "love thy neighbor," then we should convert all the church buildings into local charity centers. Turn them into food pantries, soup kitchens, homeless shelters, senior housing, and community gathering centers. Let them instruct children in morality, manners, ethics, and civics (without any of the religion/God nonsense) since the public schools aren't doing it. Turn the parochial schools into private, locally owned schools, daycare centers, or housing for the poor.

Sell off the bishops' mansions. Sell and turn the Catholic colleges and hospitals into private institutions. Use the proceeds to help the poor.

Isn't this what Jesus would do?

And, while you're at it, turn Christmas and Easter back to what they original were: winter and spring celebrations.

Turn the Catholic Church into "The Charities of Jesus" or "Catholic Charities." (The pompous heads of the Church could still exert influence as managers of the charity and make a lot of money individually, as do the leaders of other charities like the Red Cross, Boy Scouts, Anti-Defamation League, Heritage Foundation,

International Fellowship of Christians and Jews, United Way, and Wildlife Conservation Society, to name a few who take seven-figure salaries from their charities.)[83]

In Mark 10:21, Jesus says to the rich man, "Sell all that you have and give to the poor."

> "The world is my country, all mankind are my brethren, and to do good is my religion."
>
> —Thomas Paine (1737–1809), political activist

> "The good life is one inspired by love and guided by knowledge."
>
> —Bertrand Russell (1872–1970), philosopher

> "Lighthouses are more helpful than churches."
>
> —Benjamin Franklin (1706–1790), founding father

Chapter 33

Good & Evil/Heaven & Hell/Eternity

Good & Evil

Is a crocodile that savagely kills a baby wildebeest evil?

Is a spider that traps and paralyzes other insects evil?

Is a tiger that rips apart a young boy in India evil?

We wouldn't call any of the above evil. But if we substitute the word *man* in place of *crocodile, spider,* and *tiger*, each action now becomes evil to us. So, only humans can be good or evil—animals (as far as we currently know) cannot be either. Who defines what is good and what is evil? Did God define good and evil? There are examples of good and evil in the Bible, but some definitions have changed over time. For example, the Bible condones slavery, which is now condemned by most people. In the Bible, women, in general, are treated as second-class citizens.

Hundreds of years ago, the Church killed heretics—an act that they thought was good. Today, women in the Middle East and many other countries have their rights and freedoms controlled and restricted. We in the West find these actions reprehensible.

Dogfighting in some cultures is acceptable; in others, it is wrong.

In some societies, the death penalty is accepted; in some, it is not.

Some countries view abortion as murder; some see it as a woman's right to be in control of her body.

In some places, homosexuality is punishable; in some places, it's embraced.

We think that the carnage of the ancient Roman Colloseum as barbaric; in some distant future, perhaps football will be viewed the same way!

So the definition of "good" and "evil" doesn't come from God. It is defined by humans; good and evil doesn't need a God.

Heaven & Hell

Because we can think abstractly, can plan, and realize that eventually we will die, we think that it would really be nice if there were a Heaven up in the clouds, on a different level, in another type of existence, full of happiness and joy.

The Bad News: There is no Heaven after death; there is no afterlife or reincarnation.

The Okay News: There is no Hell after death.

The Reality: Heaven and Hell are right here and now. Some days are Heaven; some days are Hell. When your kids are happy, it's Heaven; when they're not, it's Hell. When you play golf with your friends, it's Heaven; when you play badly, it's Hell.

So, Heaven is the lure, the promise, the Ponzi payoff that religions promise you in exchange for influence and control over you and your money. (Would the suicidal lunatics who drove planes into the World Trade Center have died so easily if they didn't believe in Heaven (and seventy-two virgins)? One of the Boston Marathon bombers told a friend that "it's good to be a martyr and die with a smile on your face and go straight to Heaven.")

Without Heaven, people would hopefully live each day fuller.

Eternity

The Good News: There is Eternity . . . sort of.

Your eternity lies in your genes that you pass on to your children, grandchildren, and so forth. As long as your offspring have children, your genetic information survives. This is your "reincarnation." (Perhaps one day, one of your great-great-great-great-grandchildren will resemble you and exhibit many of your traits and mannerisms.) In the same manner, you are your parents, grandparents and all your ancestors. Their genetic codes live in you.

(Someday we may live on in other forms—perhaps our memories or brain functions will be able to be transferred to some sort of artificial life, another form of eternity.)

> "Of all the tyrannies that affect mankind, tyranny of religion is the worst; every other species of tyranny is limited to the world we live in; but this attempts to stride beyond the grave, and seeks to pursue us into eternity."
>
> —Thomas Paine (1737–1809), political activist

> "In those parts of the world where learning and science have prevailed, miracles have ceased; but in those parts

that are barbarous and ignorant, miracles are still in vogue."

—Ethan Allen (1738–1789), American revolutionary

"I believe that when I am dead, I am dead. I believe that with my death I am just as much obliterated as the last mosquito you and I squashed."

—Jack London (1876–1916), author

"I don't believe in Heaven and Hell. I don't know if I believe in God. All I know is that as an individual, I won't allow this life—the only thing I know to exist—to be wasted."

—George Clooney (1961–), actor

"I cannot believe in the immortality of the soul. No, all this talk of an existence for us . . . beyond the grave is wrong. It is born of our tenacity of life, our desire to go on living—our dread of coming to an end."

—Thomas A. Edison (1847–1931), inventor

"I do not fear death. I had been dead for billions and billions of years before I was born, and had not suffered the slightest inconvenience from it."

—Mark Twain (1835–1910), author

"In the fullness of time, educated people will believe that there is no soul independent of the body, and hence no life after death."

—Francis Crick (1916–2014), biologist

Chapter 34

Purpose

Most and maybe all people want to believe that their life, their existence, has purpose, has meaning.

The Bad News

From the universe's point of view, you have no purpose. You're born, you live in the universe for a brief instant of time, and you die. Eventually, your planet dies and your star (the sun) dies. That's it.

From Mother Nature's point of view, your purpose is to procreate, to nurture, to raise your offspring. What is an oak tree's purpose, a horse's purpose, a fish's purpose, a butterfly's purpose? Their purpose like ours, and all forms of life is to propagate and carry on the species.

The Good News

The good news is that you get to define your purpose.

If you want to be a great mother, be a great mother; a great father, be a great father; a great teacher, be a great teacher. Be whatever you want to be—nurse, therapist, stockbroker, clerk, farmer, politician, engineer, salesperson, analyst, coach, artist, electrician, clown, athlete, rocket scientist.

If you want to be lazy and coast, then just take it easy. If you want to be a scam artist, become a preacher. Your life's endeavor is up to you.

No outside agency, universe, or God is determining your purpose. Some think that "things happen for a reason" which, of course, doesn't make sense—things happen because they happen, with or without any specific cause. Recently, researchers from the University of Michigan and the University of Wisconsin have discovered that people who have a purpose in life—"a sense of direction and goals to live for"—are likely to be more healthy and live longer lives.[84]

So, if you haven't already, begin today to decide what will be your life's purpose.

"Man . . . is free to work out his own destiny. The responsibility is his, and so is the opportunity."

—Bertrand Russell (1872–1970), philosopher

"The more the universe seems comprehensible, the more it also seems pointless."

—Steven Weinberg (1932–), physicist

Chapter 35

"Deus de Nonsensus"

A few thousand years ago, people believed in hundreds of gods and, compared to today, knowledge was very limited and very primitive. They believed that the earth was a fraction of its true size, flat, and at the center of the universe. If we lived then, with the very limited amount of knowledge available, we probably would have believed the same. But we don't live then—we live now.

Over the years and centuries, the discovery and spread of new knowledge decreased the need for most of the gods; knowledge obsoletes gods. (Thousands of years ago, lightning must have been magical and terrifying. While it still can be terrifying, we know that it's not magical, and we know that Zeus and Thor don't cause it.) Beliefs evolved from many gods to one while new knowledge continuously developed and spread.

Crack

Along came Galileo, who said that the earth revolves around the sun. Others pointed out that the earth is just a tiny, watery, rocky planet and not the center of the universe. This astounding development meant that humans were, after all, not very important to the universe, and the institutions (religions) that guided and controlled humans were also downgraded in the whole scheme of things—a huge blow to religion/God. (Three hundred and sixty years later, the Church apologized to Galileo for disbelieving and persecuting him.)

Wham, Wham

Then Darwin in 1859 explained that humans had evolved from the very simple basic building blocks of life over billions of years—God had not created man. Another huge blow to the Church.

One hundred and fifty years later, the Church grudgingly acknowledged that Darwin's explanation of evolution was correct. Slowly, the Church's influence is deteriorating, dwindling.

Bang, Bang, Bang

Now comes Stephen Hawking, Leonard Mlodinow, Lawrence Krauss, Victor Stenger, Robert Matthews, and other noted physicists saying that research and development over the last several decades since the publication of the big bang theory in 1931, combined with growing knowledge of quantum physics (the science of infinitely small matter) lead them to believe that we don't need a God to explain the existence of the universe. The universe could come into being all by itself from nothing (which is hard for most of us to grasp at this point in time).

Now, what should the Church say? How about: "Humma . . . humma . . . humma . . . ah . . . let's see . . . well . . . even though for the last two thousand years years we instructed you to believe in the Bible literally and would persecute and kill you if you didn't, we've changed our mind. Now you should believe in the Bible only figuratively, and while humans may have evolved according to Darwin's theory, it was God who ultimately created the conditions that allowed the big bang and evolution to be able to occur."

No Proof

Over time, the Church lost its political and military powers and in 1870 lost the last of its land territory in Italy, prompting it to institute the policy of papal infallibility in an attempt to retain "spiritual power" over the people.

Now the Church is really backed into a corner by Galileo, Darwin, Hawking et al. (Pope Francis in 2014 acknowledged that the big bang and evolution were probably true—but of course still insisted that "God was still the creator.")

Since we don't need a God anymore to explain the existence of all things, we can now say that we have no evidence—no proof of a God. We may not, at this point in time, know exactly how things came to be, but we no longer need to say and believe that God did it. So, now the Church is slowly losing its last remnant of influence—its "spiritual power."

Erosion

As knowledge has spread, people have gradually embraced the fact that the earth is 4.5 billion years old, not six thousand years old (as determined from information in the Bible). More than half of the people accept the fact of evolution, that God didn't create humans. This understanding will continue to grow and be accepted as knowledge is spread—it takes time. Most people know that genes are passed from one generation to the next, helping determine who we are. Slightly under half of the people accept the big bang; in time, virtually all people will accept it.

While most people in the United States believe in God, that belief is eroding. As more knowledge is spread, understood, and accepted, the belief in religion/God reduces. That explains why the Church's membership, attendance and priests are dramatically declining in the United States, but growing in the Third World.

In olden times, the Church was the source of information. Today we have many sources: TV, Internet, newspapers, books, magazines, Twitter, Facebook, etc. and don't need the Church to tell us how to think.

In the Third World, the Church/religion may still be a main source of information, and while it took hundreds of years for knowledge to spread in the Western world, it will spread much quicker in the developing world due to the spread and availability of the Internet. Religion and God will dramatically decline; nonbelief will grow exponentially.

The mind and consciousness is not the mythical "soul." Just as we have no proof of God, we have no proof of a soul. "Faith" is a nonsensical term used to cover ignorance and lack of courage. "Faith" cannot make the earth flat, the sun revolve around the earth, 2 + 3 = 12, or account for a nonexistent God. "Faith" is putting your head in the sand, it's wishful thinking, it's delusional hoping. We're animals evolved from apes. If we have a "soul," so do they. They don't, and neither do we.

Human Conceit and Human Ignorance

The incredible vastness of the universe renders us so totally insignificant that any belief that we have significance is not believable and senseless.

Our belief that a God of all things cares individually about us demonstrates our gullibility, our ignorance, our conceit, and our susceptibility to the phony "preachers of God." (In the long run the "die is cast," for these phony, swindling preachers who will do their utmost to protect their cash flow and influence, all the while quoting from the Bible.)

It's Time

We've been programmed from birth to believe in religion/God. It's time to erase the programming. It's time to rid ourselves of the myths made up by old men thousands of years ago; men who weren't capable of figuring out things that we consider simple and basic—physical things right in front of their eyes.

It's time to end the tyranny, prejudice, and violence spread by religion/God.

It's time to grow and take responsibility for our own lives—to be empowered.

It's time to believe in ourselves and our fellow humans.

No higher power is guiding us; Heaven and Hell are right here on Earth today. It's time to pull back the wizard's curtain and reveal that the real god is ourselves; the "Higher Power" is you.

Cracks are appearing in the Church's foundation; its dominoes are lined up to fall. You and your friends need to gently push and bring about the collapse of these superstitions and myths. Time to become a "nonbeliever."

There is no Santa; there is no God.

So, our search has made us realize, made us understand, made us conclude that belief in God doesn't make sense, and as only the Church could say, all we have found in our search is a:

"Deus de Nonsensus"
"God of Nonsense"

"How can you prove that a god does not exist? . . . it is scarcely necessary to disprove what has never been proved."

—David A. Spitz (1916–1975), professor

"What can be asserted without proof can be dismissed without proof."

—Christopher Hitchens (1949–2011), author

"The biggest advantage to believing in God is you don't have to understand anything, no physics, no biology."

—James Watson (1928–), biologist

"It is said that men may not be the dreams of the gods, but rather that the gods are the dreams of men."

—Carl Sagan (1934–1996), astrophysicist

"Which is it, is man one of God's blunders or is God one of man's?"

—Friedrich Nietzsche (1844–1900), German philosopher

"If cows and horses had hands and could draw, cows would draw gods that look like cows and horses would draw gods that look like horses."

—Xenophanes (570 BC–475 BC), Greek philosopher

"Faith is nothing more than the license religious people give themselves to keep believing when reasons fail."

—Sam Harris (1967–), neuroscientist

"All thinking men are atheists."

—Ernest Hemingway (1899–1961), author

"It is among men of genius and science that atheism alone is found."

—Percy Bysshe Shelley (1792–1822), poet

"When did I realize I was God? Well, I was praying and I suddenly realized I was talking to myself."

—Peter O'Toole (1932–2013), actor

"I don't believe in God, because I don't believe in Mother Goose."

—Clarence Darrow (1857–1938), lawyer

"We must question the story logic of having an all-knowing, all-powerful God, who creates faulty humans, and then blames them for his own mistakes."

—Gene Roddenberry (1921–1991), *Star Trek* creator

"The invisible and the nonexistent look very much alike."

—John Stuart Mill (1806–1873), British philosopher

"What's God? Well, you know, when you want something really bad and you close your eyes and you wish for it? God's the guy that ignores you."

—Steve Buscemi (1957–), actor

"One of the greatest gifts science has brought to the world is continuing elimination of the supernatural."

—James Watson (1928–), biologist

"The way to see by faith is to shut the eye of reason."

—Benjamin Franklin (1706–1790), founding father

"Heaven? Well . . . nobody's actually ever come back from there, have they?"

—Ellen Nicholson (1872–1947), my Nana

Epilogue

Some Like It Hot is a classic movie comedy with Marilyn Monroe, Tony Curtis, Jack Lemon, and Joe E. Brown. Jack Lemon, a musician, dressed as a woman (Daphne) in order to avoid some mobsters and play in an all-girl band, removes his long blonde wig and reveals to Osgood Fielding III (Joe E. Brown), who wants to marry Daphne, that he's not a girl after all—he's a guy. Joe E. Brown, taking things in stride, looks at Lemon, thinks, pauses, and says, "Well, nobody's perfect."

That's the problem. Some may read this book, logically agree with it, but then emotionally say, "I still believe there's something."

Even though they understand that the current religions/God were made up by men thousands of years ago, they may think, "Well, no religion/God is perfect."

People won't give up their God beliefs overnight. It will take them weeks, months, or years to slowly come to the conclusion that there probably is no God—it will take time to change the programming.

It's up to you to start the discussion. It's up to you to change lives, hopes, expectations, and reality.

Notes

Part I

1. John Westerdall Bowker, *The Oxford Dictionary of World Religions,* Harvard University Press (1987).

2. Peter Stearns; William Langer, *The Encyclopedia of World History,* Houghton Mifflin (2001).

3. Richard Wilkinson, *The Complete Gods and Goddesses of Ancient Egypt,* Thames and Hudson (2003).

4. George Hart, *A Dictionary of Egyptian Gods and Goddesses,* Routledge (1986).

5. www.gods-and-monsters.com/egyptian-gods-and-goddesses (2014).

6. Anna Mancine, *Maat Revealed: Philosophy of Justice in Ancient Egypt,* Buenos Books America (2004).

7. www.medicalnewstoday.com/info/medicine/ancient-greek-medicine.

8. Walter Burkert, *Greek Religion,* Harvard University Press (1987).

9. Robert Shilling, *Roman Gods,* University of Chicago Press (1992).

10. www.medicalnewstoday.com/info/medicine/ancient-roman-medicine.

11. Peter Heather, *The Fall of the Roman Empire,* Oxford University Press (2005).

12. Rudolf Simek, *Dictionary of Northern Mythology,* D. S. Brewer (2007).

13. J. M. Bennett; C. W. Hollister, *Medieval Europe: A Short History,* McGraw-Hill (2006).

14. J. P. Byrne, *The Black Death,* Greenwood Publishing (2004).

15. Julie Anderson; Emm Barnes; Enna Shackleton, *The Art of Medicine,* Ilex Press (2013).

16. Lance Day; Ian McNeil, *Biographical Dictionary of the History of Technology,* Routledge (1996).

17. Roy Porter, *The Cambridge History of Medicine,* Cambridge University Press (2006).

18. www.datesandevents.org.

19. David Staley, *Encyclopedia of the History of Invention and Technology,* Facts on File (2011).

20. www2.sptimes.com/Egypt.

21. David P. Silverman, *Ancient Egypt,* Oxford University Press (2003).

22. A. G. McDowell, *Village Life in Ancient Egypt,* Harvard University Press (1993).

23. www.abyss.uoregon.edu/~js/ast123/lectures/lec01.

24. www.resafim.org.il/ad/egypt/farming.

25. www.webpages.uidaho.edu/ngier/gre13.

26. www.en.wikipedia.org/wiki/flat_earth.

27. www.touregypt.net/featurestories/prices.

28. Walter Scheidel, *Death on the Nile,* Brill (2001).

29. John F. Nunn, *Ancient Egyptian Medicine,* University of Oklahoma Press (2002).

30. www.medicalnewstoday.com/info/medicine/ancient-egyptian-medicine.

31. Bruno Halioua, Bernard Ziskind, M. B. DeBevoise, *Medicine in the Days of the Pharaohs,* Harvard University Press (2005).

32. www.reshafim.org.il/ad/egypt/religion.

33. historylink101. com/n/Egypt_1/religion_role_of_priest.

34. Jean-Marie Kruchten, "Oracles" (2001).

35. www.cyrstallinks.com/egyptpriests.

36. Dominic Montserrat, *Akhenaten: History, Fantasy, and Ancient Egypt,* Routledge (2000).

37. Ahmed Osman, *Moses and Akhenaten: The Secret History of Egypt at the Time of the Exodus,* Bear (2002).

38. Jan Assmann, *From Akhenaten to Moses,* The American University in Cairo Press (2014).

39. www.pbs.org/empires/egypt/newkingdom/priests.

40. Elliot Rabin, *Understanding The Hebrew Bible: A Reader's Guide,* KTAV Publishing (2006).

41. Tom Head, *The Absolute Beginner's Guide to the Bible,* Que Publishing (2005).

42. Dilbert Royce Burkett, *An Introduction to the New Testament and the Origins of Christianity,* Cambridge University Press (2002).

43. www.deism.com/bibleproblems.

44. www.rationalwiki.org/wiki/problems_with_biblical_inerrancy.

45. www.faithskeptic.50megs.com.

46. www.mesacc.edu/bibleproblems.

47. www.infidels.org.

48. www.listverse.com/10logicalexplanationsforincrediblebibletales.

49. www.dwindlinginunbelief.blogspot.com.

50. www.en.wikipedia.org/wiki/julian_jaynes.

Part II

1. Andrew Knoll, *Life on a Young Planet: The First Three Billion Years of Life,* Princeton University Press (2004).

2. Daniel E. Koshland, Jr., "The Seven Pillars of Life," *Science Magazine* (2002).

3. Nicholas Money, *The Amoeba in The Room,* Oxford University Press (2014).

4. W. Whitman; D. Coleman; W. Wiebe, "Prokaryotes: The Unseen Majority," *Proceedings of the National Academy of Sciences* (1998).

5. K. Nealson, "Post-Viking Microbiology," *Origins of Life and Evolution of the Biosphere* (1999).

6. Martin Blaser, *Missing Microbes,* Henry Holt (2014).

7. Charles Q. Choi, "Microbes Thrive in Deepest Spot on Earth," *Live Science* (2013).

8. Jack Szostak, "The Origin of Life on Earth" (2015).

9. Jan Sapp, *Genesis: The Evolution of Biology,* Oxford University Press (2003).

10. H. Lodish; A. Berk; P. Matsudaira; C. A. Kaiser; M. Krieger; M. P. Scott; S. L. Zipursky; J. Darnell, *Molecular Cell Biology,* W. H. Freeman (2005).

11. Andrew Berry; James Watson, *DNA: The Secret of Life,* Alfred A. Knopf (2003).

12. L. Hartwell; L. Hood; M. L. Goldberg; A. E. Reynolds; L. M. Silver; R. Veres, *Genetics: From Genes to Genomes,* McGraw-Hill (2004).

13. www.gbhealthwatch.com/trait-genetics-101.

14. www.nytimes.com/2003/11/11/science/how-did-life-begin.

15. Peter Hamilton Raven; George Brooks Johnson, *Biology,* McGraw-Hill (2002).

16. Clare Milsom, Sue Rigby, *Fossils at a Glance,* John Wiley & Sons (2009).

17. Manfred Eigen; Ruthild Winkler, *Steps toward Life: A Perspective on Evolution,* Oxford University Press (1992).

18. Steven Hawking, "Life in the Universe" (2011).

19. www.bbc.co.uk/nature/history_of_the_earth.

20. www.en.wikipedia.org/wiki/timeline_of_evolutionary_history_of_life.

21. S. L. Miller, "Production of Amino Acids Under Possible Primitive Earth Conditions," *Science Magazine* (1953).

22. D. Penny; A. Poole, "The Nature of the Last Universal Common Ancestor," *Current Opinion in Genetics & Development* (1999).

23. S. F. Mason, "Origins of Biomolecular Handedness," *Nature* (1984).

24. C. B. Charlesworth; D. Charlesworth, *Evolution,* Oxford University Press (2003).

25. D. J. Futuyma, *Evolutionary Biology,* Sinauer (1979).

26. Peter J. Bowler, *Evolution: The History of an Idea,* University of California Press (2003).

27. Peter R. Grant; Rosemary B. Grant, *How and Why Species Multiply,* Princeton University Press (2008).

28. Ben Singer, "A Brief History of the Horse in America," *Canadian Geographic Magazine* (2005).

29. Ken Miller, "The Peppered Moth: An Update," Brown University (1999).

30. James Neel, "The Thrifty Genotype in 1998," *Nutrition Reviews* (2009).

31. "Scientists Find Ancient Tibetan Altitude Gene," *Boston Globe* (July 3, 2014).

32. J. Coyne, *Why Evolution Is True,* Penguin Group (2009).

33. Stephen Jay Gould, *Evolution as Fact and Theory,* Norton (1994).

34. www.koshland-science-museum.org.

35. www.eupedia.com/forum/threads/25335-percentage-of-genetic-similarity-between-humans-and-animals.

36. "What Is Extinction?" Bristol University (2012).

37. www.wikipedia.org/wiki/genetically_modified_organism.

38. Lucas John Mix, *Life in Space,* Harvard University Press (2009).

39. Enzo Gallori, "Astrochemistry and the Origin of Genetic Material" (2010).

40. Steven J. Dick, *The Biological Universe,* Cambridge University Press (1996).

41. Daniel Strain, "Icy Moons of Saturn and Jupiter May Have Conditions Needed for Life" (2009).

42. Bill Nye, "Opinion: We Could Find Life on Another Planet. Do We Have the Will?" *National Geographic* (2014).

43. Dennis Overbye, "Far-Off Planets Like the Earth Dot the Galaxy," *New York Times* (2013).

44. Brad Steiger; John White, *Other Worlds, Other Universes,* Health Research Books (1986).

45. David Filkin; Stephen Hawking, *Stephen Hawking's Universe,* Basic Books (1998).

46. Avi Loeb, "Much-Discussed Views That Go Way Back," *New York Times* (2014).

47. www.chemistry.about.com.

48. Michael Onken, www.madsci.org (2000).

49. www.academic.brooklyn.cuny.edu/physics/sobel/nuphys/atom.

50. S. Weinberg, *The First Three Minutes: A Modern View of the Origin of the Universe,* Basic Books (1993).

51. A. H. Guth, *The Inflationary Universe: Quest for a New Theory of Cosmic Origins,* Vintage Books (1998).

52. M. Woolfson, *Time, Space, Stars and Man: The Story of the Big Bang,* World Scientific Publishing (2013).

53. S. Singh, *Big Bang: The Origins of the Universe,* Fourth Estate (2004).

54. www.age-of-the-sage.org/scientist/stephen_hawking_god_religion.

55. Alexis C. Madrigal, "A Majority of Americans Still Aren't Sure About the Big Bang," *The Atlantic* (2014).

Part III

1.www.ic.galegroup.com/ic/whtc/reference/detailspage/referencedetails.

2. www.en.wikipedia.org/wiki/2013_united_states_federal-budget.

3. www.wikipedia.org/wiki/history_of_the_earth.

4. James Trefil, *Space Atlas,* National Geographic (2012).

5. "How Many Solar System Bodies?" Solar System Dynamics (2013).

6. Kenneth J. H. Phillips, *Guide to the Sun,* Cambridge University Press (1995).

7. NASA, "The Solar Superstorm of July 2012."

8. "What Is A Galaxy? How Many Stars in a Galaxy?/the Universe?" Royal Greenwich Observatory (2006).

9. *Milky Way,* Oxford University Press (2012).

10. "How Big Is Our Universe: How Far Is It across the Milky Way?" NASA-Smithsonian (2013).

11. Aaron Robotham, Royal Astronomical Society (2014).

12. H. Shapley; H. D. Curtis, "The Scale of the Universe," National Research Council (1921).

13. www.schickworld.com/astronomy/text/distance.

14. Sara Seager, MIT; Kevin Hand, NASA, "We Are Not Alone in the Universe" (2014).

15. Bernard Carr, *Universe or Multiverse,* Cambridge University Press (2007).

Part IV

1. Dr. Adel Elsaie, "Please Revise the Bible, Again," www.us/islam.org/21 egypt

2. www.masters-table.org/pagan/sun1

3. www.kotipetripaavola.com/mitreofpope

4. www.amazingdiscoveries.org/S-deception_paganism_Catholic_mass_silver

5. www.reshafim.org.il/ad/egypt/religion/magic.htm

6. www.jesusneverexisted.com

7. www.wikipedia.org/wiki/ancient_egyptian_religion

8. Jahi Issa; Salim Faraji, *The Origin of the Word Amen,* Amen-Ra Theological Seminary Press (2006).

9. Carole M. Cusack, *The Return of the Goddess: Mythology, Witchcraft, and Feminist Spirituality,* Brill (2008).

10. *Merriam-Webster's Collegiate Encyclopedia,* Merriam-Webster (2000), p. 1211.

11. Nicholas Rogers, *Halloween: From Pagan Ritual to Party Night,* Oxford University Press (2002).

12. Anthony Aveni, *May Day: A Collision of Forces,* Oxford University Press (2004).

13. www.truthbeknown.com/victims.

14. Edward Burman, *The Inquisition: The Hammer of Heresy,* Sutton (2004).

15. www.heretication.info/_heretics

16. www.articles.exchristian.net/2002/10/how_many_people_have_been_killed_ by_christians_since_biblical_times

17. www.jewishvirtuallibrary.org/jsource/history/inquisition.

18. Wolfgang Behringer, *Witches and Witch Hunts,* Polity Press (2004).

19. Robert Chazan, "In the Year 1096: The First Crusade and the Jews" (1996).

20. www.religioustolerance.org/jud_pers.

21. Norman Housley, *Contesting the Crusades,* Blackwell Publishing (2006).

22. Alfred J. Andrea, *Encyclopedia of the Crusades* (2003).

23. Jonathan Riley-Smith, *The Crusades, Christianity, and Islam,* Columbia University Press (2011).

24. Thomas F. Madden, *The Crusades: The Essential Readings,* Blackwell (2002).

25. Kenneth Setton, *A History of the Crusades,* University of Wisconsin Press (1989).

26. James Meeker Ludow, *The Age of the Crusades,* Christian Literature Company (1896).

27. www.en.wikipedia.org/wiki/list_of_wars_and_anthropogenic_disasters_by_death_toll

28. Mack P. Holt, *The French Wars of Religion,* Cambridge University Press (2005).

29. Peter H. Wilson, *Europe's Tragedy: A History of the Thirty Years' War,* Allen Lane (2009).

30. www.religioustolerance.org/genocide5.

31. Junius Rodriguez, *The Historical Encyclopedia of World Slavery,* ABC-CLIO (1997).

32. www.theguardian.com/world/2008/feb/11/australia, "Snatched from Home for a Racist Ideal."

33. www.dailymail.co.uk/news/article-2273961/ireland-says-sorry-to-the-10,000-women-slaves.

34. Robert McClory, *Power and the Papacy: The People and Politics Behind the Doctrine of Infallibility,* Triumph Books (1997).

35. Roger Collins, *Keepers of the Keys: A History of the Papacy,* Basic Books (2009).

36. Eamon Duffy, *Saints & Sinners,* Yale University Press (2006).

37. Claudio Rendina, *The Popes: Histories and Secrets,* Seven Locks Press (2002).

38. E. R. Chamberlin, *The Bad Popes,* Sultan Publishing (2003).

39. Matthew Bunson, *The Pope Encyclopedia: an A to Z of the Holy See,* Crown Trade Paperbacks (1995).

40. www.ftarchives.net/foote/crimesofthepopes.

41. www.one-evil.org/content/people_20c_pius_xii.

42. Jonathan Gorsky, "Pius XII and the Holocaust," www.yadvashm.org.

43. Michael Phayer, *The Catholic Church and the Holocaust,* Indiana University Press (2000).

44. www.jewishvirtuallibrary.org/jsource/anti-semitism/pius.

45. John Cornwell, *Hitler's Pope: The Secret History of Pius XII,* Viking (1999).

46. Daniel Goldhagen, *A Moral Reckoning: The Role of the Catholic Church in The Holocaust and Its Unfulfilled Duty of Repair,* Little Brown (2002).

47. Carlo Falconi, *The Silence of Pius XII,* Little Brown (1970).

48. Mark Aarons; John Softus, *Unholy Trinity,* St. Martin's Press (1991).

49. www.nytimes.com/2005/01/09/world/09vatican.

50. Michael Phayer, *Pius XII, The Holocaust, and the Cold War,* Indiana University Press (2008).

51. Gerald Steinacher, *Nazis on the Run: How Hitler's Henchmen Fled Justice,* Oxford University Press (2012).

52. www.catholicherald.co.uk/news/2014/05/27/pope-francis-pius-xii-cause-for-beatification-has-stalled.

53. Jake Highton, www.open.salon.com/blog/jake_highton/2014

54. Jason Berry; Gerald Renner, *Vows of Silence: The Abuse of Power in the Papacy of John Paul II,* Free Press (2004).

55. Jonathan Levy, *The Vatican Bank,* Disinformation Press (2002).

56. www.en.wikipedia.org/wiki/institute_for_the_works_of_religion (2014).

57. www.nytimes.com/2014/06/14/opinion/cleaning-up-the-vatican-bank.

58. David A. Yallop, *In God's Name: an Investigation into the Murder of Pope John Paul I,* Robinson (2012).

59. Charles Raw, *The Moneychangers: How the Vatican Bank Enabled Robert Calvi to Steal 250 Billion Dollars,* Harvell Press (1992).

60. David Goldenberg, "10 Secrets of the Vatican Exposed," *The Week* (2013).

61. Sonya Sandage, "Vatican Bank Scandal—JPMorgan, HSBC Cease Doing Business with the Vatican" (2013).

62. Barbara Latza Nadeau, "The Money-Laundering Vatican Bank Comes Clean," *The Daily Beast* (2014).

63. M. L. Hildago, *Sexual Abuse and the Culture of Catholicism: How Priests and Nuns Became Perpetrators,* Routledge (2007).

64. Philip Jenkins, *Pedophiles and Priests: Anatomy of a Contemporary Crisis,* Oxford University Press (2001).

65. J. T. Chibnall; A. Wolf; P. N. Duckro, "A National Survey of the Sexual Trauma Experiences of Catholic Nuns," *Review of Religious Research* (1998).

66. David Ranan, *Double Cross: The Code of the Catholic Church,* Theo Press (2007).

67. www.en.wikipedia.org/wiki/arnaldus_de_villa_nova

68. www.medicalnewstoday.com/info/medicine/medieval-renaissance-medicine

69. Richard Carrier, "Christianity Was Not Responsible for Modern Science" (2010).

70. www.en.wikipedia.org/wiki/ramon_llull

71. www.en.wikipedia.org/wiki/andreas_vesalius

72. www.wired.com/2012/06/famous_persecuted_scientists

73. www.en.wikipedia.org/wiki/tommaso_campanella

74. www.scienceworld.wolfram.com/biography/bruno

75. www.markhumphreys.com/science.religion

76. www.studentpulse.com/articles/533/copernicus-galileo-and-the-church

77. www.evolution.berkeley.edu/georges_louisleclerc

78. www.rejectionofpascalswager.net/medicine

79.www.en.wikipedia.org/wiki/reactions_to_on_the_origin_of_species

80. www.en.wikipedia.org/wiki/catholic_church_and_science

Part V

1. Jennifer Viegas, "Female Chimps Seen Making, Wielding Spears," www.news.discovery.com (2015).

2. www.en.wikipedia.org/wiki/social_animal.

3. G. S. Wilkinson, "The Social Organization of the Common Vampire Bat," *Behavioral Ecology and Sociobiology* (1985).

4. Caitrin Nichol, "Do Elephants Have Souls?" *New Atlantis Magazine* (2013).

5. www.time.com/118633/ant_intelligence_google.

6.www.scientificamerican.com/article/secrets-of-ants-rafts-revealed.

7. www.csmonitor.com/science/2014/0724/dogs-display-signs-of-jealousy.

8. www.sciencerecorder.com/news/your-dog-actually does-listen-to-you.

9. Adam Steiner; David Redish, “Behavioral and Neurophysiological Correlates of Regret in Rat Decision Making,” *Nature Neuroscience Magazine* (2014).

10. www.news.sky.com/story/1310819/octopus-mum-guards-eggs.

11. Frans de Waal, *The Bonobo and The Atheist,* Norton (2013).

12. Michael Gazzaniga, *Tales from Both Sides of the Brain,* Ecco (2015).

13. www.natureworldnews.com/articles/9307/20141002/sharks-have-personalities-too.

14. www.en.wikipedia.org/wiki/separation_of_church_and_state

Part VI

1. www.gallup.com/poll/147887/americans-continue-to-believe-god

2. www.beliefnet.com/faiths/2006/08/who-believes-in-god-and-why

3.www.atheism.about.com/od/theismtheists/tp/whytheistsbelieveingod

4.www.wiki.ironchariots.org/index.php?title=50_reasons_to_believe_in_god

5. www.nytimes.com/2007/03/04/magazine/04evolution

6. www.buddha.net/e-learning/quanda03

7.www.theguardian.com/world/2005/oct13/religion.scienceandnature

8. Stanley Milgram, *Obedience to Authority: an Experimental View,* Harper Collins (1974).

9. Sigmund Freud, *The Future of an Illusion,* Martino Fine Books (2011).

10. Richard Dawkins, *The God Delusion,* Transworld Publishers (2006).

11. David J. Hanson, "Are 12-Step Programs Helpful for Alcoholism and Addictions?" (2015).

12. Peter J. McDonald; Lori I. Furbush, "Taking Different Steps" (2011).

Part VII

1. www.moroni10. com/mormon/bible_references/seeing_god.

2. www.carm.org/bible-difficulties/genesis-deuteronomy_has_ anyone_seen_god.

3. www.en.wikipedia.org/wiki/attributes_of_god.

4. www.sundayschoolresources.com/bdqualitiesofgod.

5. www.en.wikipedia.org/wiki/miracle.

6. R. C. Byrd, "Positive Therapeutic Effects of Intercessory Prayer in a Coronary Care Unit Population," *Southern Medical Journal* (1988).

7. J. M. Aviles, S. E. Whelan, D. A. Hernke, et al., "Intercessory Prayer and Cardiovascular Disease Progression in A Coronary Care Unit Population: A Randomized Controlled Trial," *Mayo Clinic Proceedings* (2001).

8. "Results of First Multicenter Trial of Intercessory Prayer, Healing Touch in Heart Patients," *Duke Medicine News* (2005).

9. H. Benson; J. A. Dusek; J. B. Sherwood; et al., "Study of the Therapeutic Effects of Intercessory Prayer in Cardiac Bypass Patients: A Multicenter Randomized Trial of Uncertainty of Receiving Intercessory Prayer," *American Heart Journal* (2006).

10. www.en.wikipedia.org/wiki/virgin_birth_of_jesus.

11. Paula Fredricksen, *From Jesus to Christ: The Origins of the New Testament Images of Jesus,* Yale University Press (2008).

12. Larry Hurtado, *Lord Jesus Christ: Devotion to Jesus in Earliest Christianity,* Eerdmans (2010).

13. www.en.wikipedia.org/wiki/miraculous_births.

14. Candida Moss, "The Strange Science behind Virgin Births," *The Daily Beast* (2014).

15. Susan Donaldson James, "Strange Nativities: Scientists Find 45 'Virgin Births,'" *British Journal of Medicine* (2013).

16. Bart Ehrman, *How Jesus Became God,* Harper One (2014).

17. Robert McClory, *Power of the Papacy: The People and Politics Behind the Doctrine of Infallibility,* Triumph Books (1997).

18. www.en.wikipedia.org/wiki/papal_infallibility.

19. Nick Pisa, "The Pope Ends State of Limbo after 800 Years," *Telegraph Media* (2007).

20. www.jewfaq.org/613.

21. www.jewfaq.org/kashrut.

22. www.chabad.org/library/article_cdo/aid/756399/jewish/the-613-commandments.

23. Ronald L. Eisenberg, *The 613 Mitzvot: A Contemporary Guide to the Commandments of Judaism,* Schreiber Publishing (2005).

24. www.en.wikipedia/wiki/ten_commandments.

25. www.medicalnewstoday.com/info/medicine/ancient-egyptian-medicine.

26. www.bbc.co.uk/history/ancient/egyptians/hearth_01.

27. www.westernseminary.edu/transformedblog/2013/10/28/hearth-and-healthcare-in-ancient-israel.

28. www.healthguidance.org/entry/6309/1/ancient-hebrew-medicine.

29. David Oshinsky, "The Last Epidemic," *Wall Street Journal* (2014).

30. Randall Haekley, "Lack of Toilets Seen As Deadly to Millions of Young Children," *Bloomberg News* (2014).

31. Gautam Naik, "Global Life Expectancy Rises by Six Years," *Wall Street Journal* (2014).

32. "Global Diarrhea Burden," Centers for Disease Control and Prevention (2014).

33. www.en.wikipedia.org/wiki/church_attendance.

34. www.pewreserach.org/fact-tank/2013/09/13/what-surveys-say-about-worship-attendance.

35. Frank Newport, "In US, Four in Ten Report Attending Church Last Week," Gallup (2013).

36. Rebecca Barnes; Lindy Lowry, "Seven Startling Facts: An Up-Close Look at Church Attendance in America" (2014).

37. www.cara.georgetown.edu/caraservices/requestedchurchstats

38. www.hirr.hartsem.edu/research/fast_facts

39. www.en.wikipedia.org/wiki/irreligion_in_the_united_states

40. www.en.wikipedia.org/wiki/list_of_ponzi_schemes

41. Mitchell Zuckoff, *Ponzi's Scheme: The True Story of A Financial Legend,"* Random House (2005).

42. Keith Veronese, "The Ladies' Deposit: Created by Women for Ripping Off Women" (2012).

43. www.voiceofsandiego.org/2009/10/13/j-david-dominelli-notorious-local-scamster

44. www.en.wikipedia/wiki/greater_ministries_international

45. Juan A. Lozano, "Stanford Gets 110 Years for Role in $7 Billion Swindle" (2012).

46. www.en.wikipedia.org/wiki/bernard_madoff

47. www.en.wikipedia.org/wiki/faith_healing

48. William Nolen, *Healing: A Doctor in Search of a Miracle*, Random House (1975).

49. www.en.wikipedia.org/wiki/kathryn_kuhlman

50. "Faith Healing," American Cancer Society (1998).

51. www.en.wikipedia.org/wiki/benny_hinn.

52. Bob McKeown, "Do You Believe in Miracles?" Canadian Broadcasting Corporation (2004).

53. Justin Peters, "Benny Hinn and Healing," CBS News (2014).

54. "Do Miracles Actually Occur?", CNN (2001).

55. www.en.wikipedia/wiki/prosperity_theology.

56. www.en.wikipedia/wiki/mother_theresa.

57. www.en.wikipedia/wiki/george_went_hensley.

58. "God Loves Uganda: How Religion Fueled an Antigay Movement," NPR (2013).

59. Bruce McClure, Deborah Byrd, "What Is a Blood Moon?" *Earth & Sky* (2014).

60. Geoff Gaherty, "Four Blood Moons: Total Lunar Eclipse Series Not a Sign of Apocalypse," *Space.com* (2014).

61. www.cnn.com/interactive/2014/08/american-archbishops-lavish-homes

62. www.theguardian.com/commentisfree/andrewbrown/2008/oct/08/ religionhowtallisgod

63. www.will-flourish.blogspot.com/2008/11/benny-hinn-false-prophet-disgrace

64. www.en.wikipedia.org/wiki/john_hagee

65. www.endevil.com/fundamentalistquotes

66. www.brainyquote.com/quotes/authors/j/jimmy_swaggart

67. www.thinkexist.com/quotes/benny_hinn/3

68. www.thinkexist.com/quotes/jim_bakker

69. www.thinkexist.com/quotes/pat_robertson

70. www.salon.com/2013/11/03/joel_osteens_gay_problem

71. www.brainyquote.com/quotes/authors/j/jerry_falwell

72. www.forwardprogressives.com/five-pat-robertson-quotes

73. www.forwardprogressives.com/bonkers-texas-preacher-says-blood-moon-is-a-sign

74. www.thinkexist.com/quotes/tammy_faye_bakker

75. www.blogs.christianpost.com/ridiculous-reads/10-crazy-quotes-from-televangelists

76. www.brainyquote.com/quotes/j/joelosteen

77. www.huffingtonpost.com/katie-halper/top-ten-hagee-isms

78. www.en.wikiquote.org/wiki/ted_haggard

79. www.rightwingwatch.org/content/john-hagee-says-ebola-gods-judgment

80. www.slate.com/blogs/xx_factor/2012/07/23/rick_warren_accusing

81. www.en.wikipedia.org/wiki/list_of_atheists

82. www.en.wikipedia.org/wiki/list_of_agnostics

83. www.charitywatch.org/hottopics/top25

84. Karen Kaplan, "Study: A Sense of Purpose Can Help Cut Costs," *Los Angeles Times* (2014).

Bibliography

Amir Aczel. *Why Science Does Not Disprove God.* William Morrow (2014).

James Allen. *"Monotheism: The Egyptian Roots." Biblical Archaeology Review* (July 1999).

Alfred J. Andrea. *Encyclopedia of the Crusades.* (2003).

Carol A.R. Andrews. *Amulets of Ancient Egypt.* University of Texas Press (1994).

The AP-GFK Poll (March 2014).

Karen Armstrong. *A History of God: The 4,000-Year Quest of Judaism. Christianity and Islam.* Ballantine Books (1994).

Enrico Ascalone. *Mesopotamia: Assyrians, Sumerians, Babylonians (Dictionaries of Civilizations).* University of California Press (2007).

Jan Assmann. *The Search for God in Ancient Egypt.* Cornell University Press (2001).

Jan Assmann. *From Akhenaten to Moses.* The American University in Cairo Press (2014).

Roger S. Bagnell; Bruce W. Frier. *The Demography of Ancient Egypt.* Cambridge University Press (1994).

Dr. Emily Baldwin. "Water, Water Everywhere." *Astronomy Now* (2012).

Randall Herbert Balmer. *Encyclopedia of Evangelicalism.* John Knox Press (2002).

Kathryn A. Bard; Stephen Blake Shubert. *Encyclopedia of the Archaeology of Ancient Egypt.* Routledge (1999).

Itzhak Bars; John Terning. *Extra Dimension in Space and Time.* Springer (2009).

Peter Bellwood. *First Farmers: The Origins of Agricultural Societies.* Blackwell Publishers (2004).

Andrew Berry; James Watson. *DNA: The Secret of Life.* Alfred A. Knopf (2003).

Peter J. Bowler. *Evolution: The History of an Idea.* University of California Press (2003).

Daniel Boyarin. *Dying for God: Martyrdom and the Making of Christianity and Judaism.* Stanford University Press (1999).

Suzanne Brown-Fleming. *The Holocaust and Catholic Conscience Cardinal Aloisius Muench and the Guilt Question in Germany.* University of Notre Dame Press (2006).

Matthew Bunson. *The Pope Encyclopedia: An A to Z of the Holy See.* Crown Trade Paperbacks (1995).

Walter Burkert. *Greek Religion.* Harvard University Press (1987).

Edward Burman. *The Inquisition: The Hammer of Heresy.* Sutton Publishers (2004).

R. M. Canup; E. Asphaug. "Origin of the Moon in a Giant Impact Near the End of the Earth's Formation." *Nature* (2001).

S. Carmona; M. Ezzamel. "Accounting and Forms of Accountability in Ancient Civilizations: Mesopotamia and Egypt" (2005).

Sean Carroll. "Dark Matter. Dark Energy: The Dark Side of the Universe." The Teaching Company (2007).

CFA.HARVARD.EDU. "How Big Is Our Universe: How Far Is It across the Milky Way?" (2013).

Eric J. Chaisson. *Cosmic Evolution.* Harvard University Press (2005).

E. R. Chamberlin. *The Bad Popes.* Dorset (1969).

D. S. Chambers. *Popes. Cardinals & War: The Military Church in Renaissance and Early Modern Europe.* Tauris (2006).

C. B. Charlesworth; D. Charlesworth. *Evolution.* Oxford University Press (2003).

Shaye J. D. Cohen. *From the Maccabees to the Mishnah.* Westminster John Knox Press (2006).

Roger Collins. *Keepers of the Keys: A History of the Papacy.* Basic Books (2009).

Frank J. Coppa. *The Life and Pontificate of Pope Pius XII: Between History and Controversy.* Catholic University of America Press (2013).

John Cornwell. *Hitler's Pope: The Secret History of Pius XII.* Viking (1999).

J. Coyne. *Why Evolution Is True.* Penguin Group (2009).

Rosalie David; Patricia Lambert-Zazulak. *Egyptian Mummies and Modern Science.* Cambridge University Press (2008).

Norman Davies. *Europe: A History.* Oxford University Press (1996).

Richard Dawkins. *The Selfish Gene*. Oxford University Press (1989).

Richard Dawkins. *The Blind Watchmaker.* W. W. Norton & Company (1996).

Richard Dawkins. *The God Delusion.* Houghton Mifflin (2006).

Lance Day; Ian McNeil. *Biographical Dictionary of the History of Technology*. Routledge (1996).

Daniel C. Dennett. *Breaking the Spell.* Viking (2006).

David Deutsch. *The Fabric of Reality.* Penguin Books (2011).

Frans de Waal. *Our Inner Ape.* Riverhead Books (2005).

Stephen J. Dick. *The Biological Universe: The Twentieth Century Extraterrestrial Life Debate and the Limits of Science.* Cambridge University Press (1996).

Alexander Drake. *The Invention of Religion.* The Emperor Has No Clothes Press (2012).

Eamon Duffy. *Saints & Sinners.* Yale University Press (2006).

Bart Ehrman. *How Jesus Became God.* Harper One (2014).

Ronald L. Eisenberg. *The 613 Mitzvot: A Contemporary Guide to the Commandments of Judaism.* Schreiber Publishing (2005).

Eso.org. "The Biggest Star in the Sky" (1997).

Evoled.dbs.umt.edu. "Teach Evolution and Make It Relevant" (2014).

R. O. Faukner. *The Ancient Egyptian Pyramid Texts.* Kessinger (2004).

Richard Fortey. *Life: A Natural History of the First Four Billion Years of Life on Earth.* Vintage Books (1999).

Paula Fredricksen. *From Jesus to Christ: The Origins of the New Testament Images of Jesus.* Yale University Press (2008).

www.Georgetown.edu/caraservices. "Church Statistics." (2014).

Daniel Goldhagen. *A Moral Reckoning: The Role of the Catholic Church in the Holocaust and Its Unfilled Duty of Repair.* Little Brown & Company (2002).

Edward Grant. *The Foundations of Modern Science in the Middle Ages: Their Religious, Institutional, and Intellectual Contexts.* Cambridge University Press (1996).

Nicholas Grimal. *A History of Ancient Egypt.* Blackwell Books (1992).

A. H. Guth. *The Inflationary Universe: Quest for a New Theory of Cosmic Origins.* Vintage Book (1998).

Bruno Halioua; Bernard Ziskind; M.B. DeBevoise. *Medicine in the Days of the Pharaohs.* Harvard University Press (2005).

James Hannam. *The Genesis of Science.* Regnery Publishing (2011).

Sam Harris. *The End of Faith.* W. W. Norton & Company (2005).

Sam Harris. *Letter to a Christian Nation.* Knopf (2006).

George Hart. *A Dictionary of Egyptian Gods and Goddesses.* Routledge (1986).

L. Hartwell; L. Hood; M. L. Goldberg; A. E. Reynolds; L. M. Silver; R. Veres. *Genetics: From Genes to Genomes.* McGraw-Hill (2004).

Stephen Hawking. *A Brief History of Time.* Bantam Books (1988).

Stephen Hawking; Leonard Mlodinow. *The Grand Design.* Bantam Books (2010).

Christopher Hitchens. *God Is Not Great.* Twelve (2007).

Ralph W. Hood; William Paul Williamson. *Them That Believe: The Power and Meaning of the Christian Serpent Handling Tradition.* University of California Press (2008).

Mack P. Holt. *The French Wars of Religion 1562–1629.* Cambridge University Press (2005).

Ian Spencer Hornsey. *A History of Beer and Brewing.* The Royal Society of Chemistry (2003).

Andrew Hunt Gordon; Calvin W. Schwabe. *The Quick and the Dead: Biomedical Theory in Ancient Egypt.* Brill (2004).

Toby E. Huff. *The Rise of Early Modern Science: Islam, China, and the West.* Cambridge University Press (2003).

Larry W. Hurtado. *Lord Jesus Christ: Devotion to Jesus in Earliest Christianity*. Eerdmans (2010).

Issa & Faraji. *The Origin of the Word Amen.* Amen Ra Theological Seminary Press (2007).

Philip Jenkins. *Pedophiles and Priests: Anatomy of a Contemporary Crisis.* Oxford University Press (2001).

Penn Jillette. *God No!: Signs You May Already Be an Atheist and Other Magical Tales.* Simon & Schuster (2011).

David Jones; Russell Woodbridge. *Health, Wealth & Happiness: Has the Prosperity Gospel Overshadowed the Gospel of Christ?.* Kreigel Publications (2011).

Neil Kagan. *Concise History of the World.* National Geographic Society (2006).

James F. Kasting; David Catling. "Evolution of a Habital Planet." *Astronomy and Astrophysics* (2003).

Lawrence Krauss. *A Universe from Nothing.* Free Press (2012).

Jonathan Levy. *The Vatican Bank.* Disinformation Press (2002).

Lise Manniche. *Sexual Life in Ancient Egypt.* Routledge (1987).

Gaston Maspero. *Popular Stories of Ancient Egypt.* Kessinger Publishing (2003).

Robert McClory. *Power and the Papacy: The People and Politics behind the Doctrine of Infallibility.* Triumph Books (1997).

A. G. McDowell. *Village Life in Ancient Egypt.* Oxford University Press (2002).

Bojana Mojsov. *Osiris: Death and Afterlife of a God.* Blackwell Publishing (2005).

Dominic Montserrat. *Akenaten: History, Fantasy, and Ancient Egypt.* Routledge (2000).

A. Morbidelli; J. Chambers; J. Lunine; J. Petit; F. Robert; G. B. Valsecchi; K. E. Cyr. "Source Regions and Timescales for the Delivery of Water to the Earth." *Meteoritics & Planetary Science* (2000).

Joseph Needham. *Science and Civilization in China.* Cambridge University Press (1987).

David S. New. *Holy War: The Rise of Militant Christian, Jewish, and Islamic Fundamentalism.* McFarland (2013).

William Nolen. *Healing: A Doctor in Search of A Miracle.* Random House (1975).

John Francis Nunn. *Ancient Egyptian Medicine.* University of Oklahoma Press (2002).

Ahmed Osman. *Moses and Akhenaten: The Secret History of Egypt at the Time of the Exodus.* Bear & Company (2002).

Francis E. Peters; John L. Esposito. *The Children of Abraham, Judaism. Christianity. Islam.* Princeton University Press (2006).

Michael Phayer. *Pope Pius XII. The Holocaust and the Cold War.* Indiana University Press (2008).

Kenneth J.H. Phillips. *Guide to the Sun.* Cambridge University Press (1995).

Clifford Pickover. *Archimedes to Hawking: Laws of Science and the Great Minds behind Them.* Oxford University Press (2008).

Roy Porter. *The Cambridge History of Medicine* (2006).

David Ranan. *Double Cross: The Code of the Catholic Church.* Theo Press (2007).

James Randi. *The Faith Healers.* Prometheus Press (1989).

Donald B. Redford. *Akhenaten: The Heretic King.* Princeton University Press (1987).

Claudio Rendina. *The Popes: Histories and Secrets.* Seven Locks Press (2002).

Reshafim.org.

H. A. Rey. *The Stars.* Houghton Mifflin Harcourt (2008).

Janet E. Richards. *Society and Death in Ancient Egypt.* Cambridge University Press (2005).

Jonathan Riley-Smith. *The Oxford History of the Crusades.* Oxford University Press (1999).

Gay Robins. *Women in Ancient Egypt.* Harvard University Press (1993).

Junius P. Rodriguez. *The Historical Encyclopedia of World Slavery.* ABC-CLIO (1997).

Royal Greenwich Observatory. "What Is a Galaxy? How Many Stars in a Galaxy/the Universe?" (2006).

Fernando Salmon; Robert E. Bjork. *The Oxford Dictionary of the Middle Ages.* Oxford University Press (2010).

Walter Scheidel. *Death on the Nile: Disease and the Demography of Roman Egypt.* Brill (2001).

Sciencedaily.com. "Nearby Planet-Forming Disk Holds Water for Thousands of Oceans" (2011).

Ian Shaw; Paul Nicholson. *Dictionary of Ancient Egypt.* British Museum Press (1995).

David P. Silverman. *Ancient Egypt.* Oxford University Press (2003).

Rudolf Simek. *Dictionary of Northern Mythology.* D. S. Brewer (2007).

S. Singh. *Big Bang: The Origins of the Universe.* Fourth Estate (2004).

Space.com. "100 Billion Alien Planets Fill Our Milky Way Galaxy" (2013).

David J. Staley. *Encyclopedia of the History of Invention and Technology.* Facts on File (2000).

Peter N. Stearns; William L. Langer. *The Encyclopedia of World History: Ancient. Medieval and Modern, Chronologically Arranged.* Houghton Mifflin (2001).

Gerald Steinacher. *Nazis on the Run: How Hitler's Henchmen Fled Justice.* Oxford University Press (2012).

Victor J. Stenger. *God and the Folly of Faith.* Prometheus Books (2012).

Claire Sterling. *Octopus: How the Long Reach of the Sicilian Mafia Controls the Global Narcotics Trade.* Simon & Schuster (1990).

James Trefil. *Space Atlas.* National Geographic Society (2012).

Colin Tudge. *Neanderthals. Bandits and Farmers: How Agriculture Really Began.* Wiedenfeld & Nicholson (1998).

William Turner. "Raymond Lully." *The Catholic Encyclopedia* (2013).

Ray Villard; Wendy L. Freedman. "Hubble Space Telescope Measures Precise Distance to the Most Remote Galaxy Yet." Hubble Site (1994).

C. V. Wedgwood. *Thirty Years' War.* NYRB Classics (2005).

S. Weinberg. *The First Three Minutes: A Modern View of the Origin of the Universe.* Basic Books (1993).

Wikipedia.org—multiple and various.

Richard H. Wilkinson. *The Complete Gods and Goddesses of Ancient Egypt.* Thames & Hudson (2003).

Peter H. Wilson. *Europe's Tragedy: A History of the Thirty Years' War.* Allen Lane (2009).

M. Woolfson. *Time, Space, Stars, and Man: The Story of the Big Bang.* World Scientific Publishing (2013).

Mitchell Zuckoff. *Ponzi's Scheme: The True Story of a Financial Legend.* Random House (2005).

Acknowledgements

Many thanks to Cheryl, Ruthie, John, Ellen, Frank, Kevin, Jen, Paul, Gayle, Joan, Michelle, Debbi, Gilbert and Mary Ellen who contributed immensely to this book.

Mom, don't be worried—I'm not going to Hell. Dad, thank you. (Besides, Mom, I've got loads of plenary indulgences from attending First Friday Masses when I was in high school. I'm "golden.")

About the Author

Jack Carlow has held influential management positions in many small and large companies. He's consulted with and motivated hundreds of businesses to grow and prosper, and, today, his firm is recognized as one of the leading small-business consulting firms in the United States. He holds a degree in engineering and an MBA.

Raised as a Catholic in a loving home with several siblings, he attended weekly Mass and confession when young. His Irish-Catholic upbringing and parochial school education gave him insight into religion and God.

As he grew older, questions and doubts emerged. He witnessed family and friends die due to lack of proper medical treatment as a result of their religious beliefs. With little active thought, things stewed in his mind, processing subconsciously, altering long-held opinions and beliefs.

Raised by his parents to think creatively, inquisitively, and independently and trained in engineering school to think logically, he began his search for the meaning of religion and God. The result is this book, which takes the reader logically and easily from belief to questioning to nonbelief.

Made in United States
Troutdale, OR
11/03/2026

56845130R00163